SpringerBriefs in Applied Sciences and Technology

Computational Intelligence

For further volumes:
http://www.springer.com/series/10618

Oscar Castillo · Patricia Melin

Recent Advances in Interval Type-2 Fuzzy Systems

Springer

Prof. Dr. Oscar Castillo
Division of Graduate Studies
Tijuana Institute of Technology
Chula Vista
CA 91909
USA

Prof. Dr. Patricia Melin
Division of Graduate Studies
Tijuana Institute of Technology
Chula Vista
CA 91909
USA

ISSN 2191-530X
e-ISSN 2191-5318
ISBN 978-3-642-28955-2
e-ISBN 978-3-642-28956-9
DOI 10.1007/978-3-642-28956-9
Springer Heidelberg New York Dordrecht London

Library of Congress Control Number: 2012935671

Printed on acid-free paper

Springer is part of Springer Science+Business Media (www.springer.com)

Preface

We describe in this book, new methods for building intelligent systems using type-2 fuzzy logic and soft computing techniques. In this book, we are extending the use of fuzzy logic to a higher order, which is called type-2 fuzzy logic. Combining type-2 fuzzy logic with traditional SC techniques, we can build powerful hybrid intelligent systems that can use the advantages that each technique offers. We consider in this book the use of type-2 fuzzy logic and traditional SC techniques to solve problems in real-world applications.

This book is intended to be a reference for scientists and engineers interested in applying type-2 fuzzy logic for solving problems in pattern recognition, intelligent control, intelligent manufacturing, robotics and automation. This book can also be used as a reference for graduate courses like the following: soft computing, intelligent pattern recognition, computer vision, applied artificial intelligence, and similar ones. We consider that this book can also be used to get novel ideas for new lines of research, or to continue the lines of research proposed by the authors of the book.

In Chap. 1, we begin by offering a brief introduction of the potential use of type-2 fuzzy logic in different real-world applications. We discuss the application of type-2 fuzzy logic in problems of pattern recognition. We also describe the use of type-2 fuzzy logic in problems of intelligent control of non-linear plants. We also outline the application of type-2 fuzzy logic in real-world applications of intelligent manufacturing, robotics and automation.

We describe in Chap. 2 the basic concepts, notation, and theory of type-2 fuzzy logic, which is a generalization of type-1 fuzzy logic. Type-2 fuzzy logic enables the management of uncertainty in a more complete way. This is due to the fact that in type-2 membership functions we also consider that there is uncertainty in the form of the functions, unlike type-1 membership functions in which the functions are considered to be fixed and not uncertain.

We describe in Chap. 3 a brief overview of the basic concepts from bio-inspired optimization methods needed for this work. In particular, the methods that are covered in this chapter are: particle swarm optimization, genetic algorithms and ant colony optimization.

We offer in Chap. 4 a representative review of the works using a bio-inspired optimization technique, like genetic algorithms (GAs), for automating the design process of type-2 fuzzy systems. This overview has the goal of providing the reader with an idea of the diversity of applications that have been achieved using genetic algorithms for type-2 fuzzy system optimization.

We describe in Chap. 5 a representative review of works on optimizing type-2 fuzzy systems using different kinds of particle swarm optimization (PSO) algorithms to illustrate the advantages of using this optimization technique for automating the design process of type-2 fuzzy systems.

We describe in Chap. 6 a representative and brief review of the works that have used ant colony optimization (ACO) to illustrate the advantages of using this optimization technique for automating the design process or parameters of type-2 fuzzy systems.

We describe in Chap. 7 some other works reported in the literature optimizing type-2 fuzzy systems using different kinds of optimization algorithms (other than GAs, PSO or ACO, which were covered in previous chapters). Most of these works have had relative success according to the different areas of application. In this chapter, we offer a representative and brief review of these types of works to illustrate the advantages of using the corresponding optimization techniques for automating the design process or parameters of type-2 fuzzy systems.

We describe in Chap. 8 as an illustration the optimization of the membership functions' parameters of an interval type-2 fuzzy logic controller in order to find the optimal intelligent controller for an autonomous wheeled mobile robot. The optimization method that was used is based on the chemical reaction paradigm. Simulation results with the chemical optimization paradigm are very good and are shown to outperform other optimization methods for the same control problem.

We describe in Chapter 9 a method for the design of a Type-2 Fuzzy Logic Controller (FLC-T2) and a Type-1 Fuzzy Logic Controller (FLC-T1) using Genetic Algorithms. The two controllers were tested with different levels of uncertainty to regulate speed in a direct current motor. The controllers were synthesized in Very High Description Language (VHDL) code for a Field Programmable Gate Array (FPGA), using the Xilinx System Generator of Xilinx ISE and Matlab-Simulink. Comparisons were made between the FLC-T1 versus FLC-T2 in VHDL code and also with a Proportional Integral Differential (PID) Controller. To evaluate the difference in performance of the three types of controllers, the t-student statistical test was used with the type-2 controller resulting to be the best one for this problem.

We describe in Chap. 10 a general overview of the area of type-2 fuzzy system optimization. Also, possible future trends that we can envision based on the review of this area are presented. It has been well-known for a long time, that designing fuzzy systems is a difficult task, and this is especially true in the case of type-2 fuzzy systems. The use of GAs, ACO and PSO in designing type-1 fuzzy systems has become a standard practice for automatically designing this sort of systems. This trend has also continued to the type-2 fuzzy systems area, which has been accounted for with the review of papers presented in the previous chapters. In this chapter a summary of the total number of papers published in the area of type-2 fuzzy system optimization is also presented, so that the increasing trend occurring in this area can be better appreciated.

We end this preface of the book by giving thanks to all the people who have help or encourage us during the writing of this book. First of all, we would like to thank our colleague and friend Prof. Janusz Kacprzyk for always supporting our work, and for motivating us to write our research work. We would also like to thank our colleagues working in Soft Computing, which are too many to mention each by their name. Of course, we need to thank our supporting agencies, CONACYT and DGEST, in our country for their help during this project. We have to thank our institution, Tijuana Institute of Technology, for always supporting our projects. Finally, we thank our families for their continuous support during the time that we spend in this project.

Mexico

Prof. Dr. Oscar Castillo
Prof. Dr. Patricia Melin

Contents

Chapter 1
Introduction

A review of the optimization methods used in the design of type-2 fuzzy systems, which are relatively novel models of imprecision, is presented in this book. The fundamental focus of the book is based on the basic reasons of the need for optimizing type-2 fuzzy systems for different areas of application. Recently, bio-inspired methods have emerged as powerful optimization algorithms for solving complex problems. In the case of designing type-2 fuzzy systems for particular applications, the use of bio-inspired optimization methods have helped in the complex task of finding the appropriate parameter values and the right structure of the fuzzy systems. In this book, we review the application of genetic algorithms, particle swarm optimization and ant colony optimization, as three different paradigms that help in the design of optimal type-2 fuzzy systems. We also provide a comparison of results for the different optimization methods for the case of designing type-2 fuzzy systems.

Uncertainty affects decision-making and emerges in a number of different forms. The concept of information is inherently associated with the concept of uncertainty [1, 2]. The most fundamental aspect of this connection is that the uncertainty involved in any problem-solving situation is a result of some information deficiency, which may be incomplete, imprecise, fragmentary, not fully reliable, vague, contradictory, or deficient in some other way. Uncertainty is an attribute of information [3]. The general framework of fuzzy reasoning allows handling much of this uncertainty and fuzzy systems employ type-1 fuzzy sets, which represent uncertainty by numbers in the range [0, 1]. When an entity is uncertain, like a measurement, it is difficult to specify its exact value, and of course a type-1 fuzzy set makes more sense than a traditional set [3, 4]. However, it is not reasonable to use an accurate membership function for something uncertain, so in this case what we need is another type of fuzzy sets, those which are able to handle these uncertainties, the so called type-2 fuzzy sets [5, 6]. The amount of uncertainty in a system can be reduced by using type-2 fuzzy logic

O. Castillo and P. Melin, *Recent Advances in Interval Type-2 Fuzzy Systems*,
SpringerBriefs in Computational Intelligence, DOI: 10.1007/978-3-642-28956-9_1,

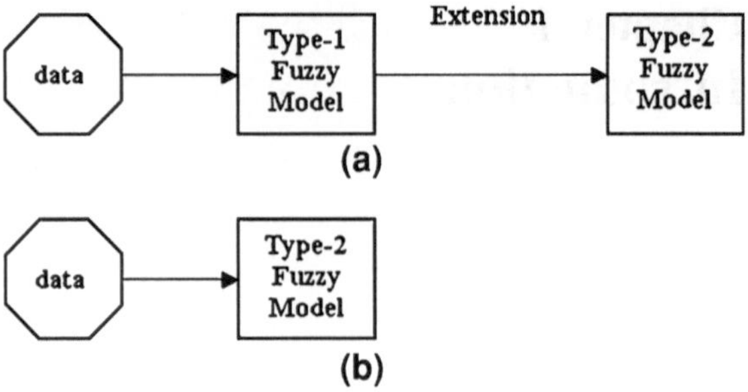

Fig. 1.1 Two categories of approaches to the design of interval type-2 fuzzy systems (models): **a** methods based on an augmentation of existing type-1 fuzzy model, and **b** methods aimed at the direct development of type-2 fuzzy models from data

because this logic offers better capabilities to handle linguistic uncertainties by modeling vagueness and unreliability of information [7, 8].

Type-2 fuzzy models have emerged as an interesting generalization of fuzzy models based upon type-1 fuzzy sets [5, 9]. There have been a number of claims put forward as to the relevance of type-2 fuzzy sets being regarded as generic building constructs of fuzzy models [10–12]. Likewise, there is a record of some experimental evidence showing some improvements in terms of accuracy of fuzzy models of type-2 over their type-1 counterparts [13–17]. Unfortunately, no systematic and comprehensive design framework has been provided and while improvements over type-1 fuzzy models were evidenced, it is not clear whether this effect could always be expected. Furthermore, it is not demonstrated whether the improvement is substantial enough and fully legitimized given the substantial optimization overhead associated with the design of this category of models. There have been a lot of applications of type-2 in intelligent control [18–25], pattern recognition [26–30], intelligent manufacturing [15, 31, 32], time series prediction [13, 33], and others [34–39]. However, no general design strategy for finding the optimal type-2 fuzzy model has been proposed, and for this reason bio-inspired algorithms have been used to try in find these optimal type-2 models.

In general, the methods for designing a type-2 fuzzy model based on experimental data can be classified into two categories as illustrated in Fig. 1.1. The first category of methods assumes that an optimal (in some sense) type-1 fuzzy model has already been designed and afterwards a type-2 fuzzy model is constructed through some sound augmentation of the existing model. The second class of design methods is concerned with the construction of the type-2 fuzzy model directly from experimental data. In both cases, an optimization method can help in obtaining the optimal type-2 fuzzy model for the particular application.

Recently, bio-inspired methods have emerged as powerful optimization algorithms for solving complex problems. In the case of designing type-2 fuzzy systems for particular applications, the use of bio-inspired optimization methods have helped in the complex task of finding the appropriate parameter values and structure of the fuzzy systems. In this book, we consider a review on the application of genetic algorithms, particle swarm optimization and ant colony optimization as three different paradigms that help in the design of optimal type-2 fuzzy systems. We also mention some hybrid approaches and other optimization methods that have been applied in the problem of designing optimal type-2 fuzzy systems in different domains of application.

The rest of the book is organized as follows. In Chap. 2 some basic definitions of type-2 fuzzy systems are presented. Chapter 3 describes some basic concepts of bio-inspired optimization. Chapter 4 describes the application of genetic algorithms for the optimization of type-2 fuzzy systems. In Chap. 5 a review of different approaches for the application of particle swarm optimization in type-2 fuzzy systems design are presented. Chapter 6 presents an overview of ant colony optimization methods applied in type-2 fuzzy systems design. Chapter 7 discusses other approaches that have been used to optimize type-2 fuzzy systems. Chapter 8 describes in detail a particular application of type-2 fuzzy systems in the control of an autonomous robot. Chapter 9 presents an overview of the area and future trends of research in optimal type-2 fuzzy system design.

References

1. P. Melin, O. Castillo, *Modelling, Simulation and Control of Non-Linear Dynamical Systems* (Taylor and Francis, London, 2002)
2. J.M. Mendel, Uncertainty, fuzzy logic, and signal processing. Sig. Process. J. **80**, 913–933 (2000)
3. L.A. Zadeh, The concept of a linguistic variable and its application to approximate reasoning. Inf. Sci. **8**, 43–80 (1975)
4. J.R. Jang, C.T. Sun, E. Mizutani, *Neuro-Fuzzy and Soft Computing* (Prentice Hall, Upper Saddle River, 1997)
5. O. Castillo, P. Melin, *Type-2 Fuzzy Logic: Theory and Applications* (Springer, Heidelberg, 2008)
6. N.N. Karnik, J.M. Mendel, *An Introduction to Type-2 Fuzzy Logic Systems*, Technical Report, (University of Southern California, 1998)
7. M. Wagenknecht, K. Hartmann, Application of fuzzy sets of type 2 to the solution of fuzzy equations systems. Fuzzy Sets Syst. **25**, 183–190 (1988)
8. R.R. Yager, Fuzzy subsets of type II in decisions. J. Cybern. **10**, 137–159 (1980)
9. H. Hagras, Hierarchical type-2 fuzzy logic control architecture for autonomous mobile robots. IEEE Trans. Fuzzy Syst. **12**, 524–539 (2004)
10. S. Coupland, R. John, New geometric inference techniques for type-2 fuzzy sets. Int. J. Approx. Reason. **49**, 198–211 (2008)
11. J.T. Starczewski, Efficient triangular type-2 fuzzy logic systems. Int. J. Approx. Reason. **50**, 799–811 (2009)
12. C. Walker, E. Walker, Sets with type-2 operations. Int. J. Approx. Reason. **50**, 63–71 (2009)
13. N.S. Bajestani, A. Zare, in *Application of Optimized Type-2 Fuzzy Time Series to Forecast Taiwan Stock Index*. Second International Conference on Computer, Control and Communication (2009), pp. 275–280
14. J.R. Castro, O. Castillo, P. Melin, A. Rodriguez-Diaz, A hybrid learning algorithm for a class of interval type-2 fuzzy neural networks. Inf. Sci. **179**, 2175–2193 (2009)
15. T. Dereli, A. Baykasoglu, K. Altun, A. Durmusoglu, I.B. Turksen, Industrial applications of type-2 fuzzy sets and systems: a concise review. Comput. Ind. **62**, 125–137 (2011)

16. C. Leal-Ramirez, O. Castillo, P. Melin, A. Rodriguez-Diaz, Simulation of the bird age-structured population growth based on an interval type-2 fuzzy cellular structure. Inf. Sci. **181**, 519–535 (2011)
17. R. Martinez, O. Castillo, L. Aguilar, Optimization with genetic algorithms of interval type-2 fuzzy logic controllers for an autonomous wheeled mobile robot: a comparison under different kinds of perturbations, in *Proceedings of the IEEE FUZZ Conference*, 2008, paper # FS0225
18. O. Castillo, P. Melin, *Soft Computing for Control of Non-Linear Dynamical Systems* (Springer, Heidelberg, 2001)
19. O. Castillo, L.T. Aguilar, N.R. Cazarez-Castro, S. Cardenas, Systematic design of a stable type-2 fuzzy logic controller. Appl. Soft Comput. J. **8**, 1274–1279 (2008)
20. M. Hsiao, T.H.S. Li, J.Z. Lee, C.H. Chao, S.H. Tsai, Design of interval type-2 fuzzy sliding-mode controller. Inf. Sci. **178**(6), 1686–1716 (2008)
21. P. Melin, O. Castillo, A new method for adaptive model-based control of non-linear dynamic plants using a neuro-fuzzy-fractal approach. J. Soft Comput. **5**, 171–177 (2001)
22. P. Melin, O. Castillo, A new method for adaptive model-based control of nonlinear plants using type-2 fuzzy logic and neural networks, in *Proceedings of the IEEE FUZZ Conference*, 2003, pp. 420–425
23. T. Ozen, J.M. Garibaldi, Investigating Adaptation in Type-2 Fuzzy Logic Systems Applied to Umbilical Acid-Base Assessment, in *European Symposium on Intelligent Technologies, Hybrid Systems and their implementation on Smart Adaptive Systems (EUNITE 2003)*, 2003, Oulu
24. R. Sepulveda, O. Castillo, P. Melin, O. Montiel, An efficient computational method to implement type-2 fuzzy logic in control applications. Adv. Soft Comput. **41**, 45–52 (2007)
25. R. Sepulveda, O. Castillo, P. Melin, A. Rodriguez-Diaz, O. Montiel, Experimental study of intelligent controllers under uncertainty using type-1 and type-2 fuzzy logic. Inf. Sci. **177**(10), 2023–2048 (2007)
26. P. Melin, O. Castillo, *Hybrid Intelligent Systems for Pattern Recognition* (Springer, Heidelberg, 2005)
27. O. Mendoza, P. Melin, O. Castillo, G. Licea, Type-2 fuzzy logic for improving training data and response integration in modular neural networks for image recognition. Lecture Notes in Artificial Intelligence, vol. 4529 (2007), pp. 604–612
28. O. Mendoza, P. Melin, O. Castillo, Interval type-2 fuzzy logic and modular neural networks for face recognition applications. Appl. Soft Comput. J. **9**, 1377–1387 (2009)
29. O. Mendoza, P. Melin, G. Licea, Interval type-2 fuzzy logic for edges detection in digital images. Int. J. Intell. Syst. **24**, 1115–1133 (2009)
30. J. Urias, D. Hidalgo, P. Melin, O. Castillo, A method for response integration in modular neural networks with type-2 fuzzy logic for biometric systems. Adv. Soft Comput. **41**, 5–15 (2007)
31. P. Melin, O. Castillo, An intelligent hybrid approach for industrial quality control combining neural networks, fuzzy logic and fractal theory. Inf. Sci. **177**, 1543–1557 (2007)
32. M.H.F. Zarandi, I.B. Turksen, O.T. Kasbi, Type-2 fuzzy modelling for desulphurization of steel process. Expert Syst. Appl. **32**, 157–171 (2007)
33. O. Castillo, P. Melin, Hybrid intelligent systems for time series prediction using neural networks, fuzzy logic and fractal theory. IEEE Trans. Neural Netw. **13**, 1395–1408 (2002)
34. L. Astudillo, O. Castillo, L.T. Aguilar, R. Martinez, Hybrid control for an autonomous wheeled mobile robot under perturbed torques. Lecture Notes in Computer Science, vol. 4529 (2007), pp. 594–603
35. J.R. Castro, O. Castillo, P. Melin, An interval type-2 fuzzy logic toolbox for control applications, in *Proceedings of FUZZ-IEEE 2007*, London, pp. 1–6
36. J.R. Castro, O. Castillo, L.G. Martinez, Interval type-2 fuzzy logic toolbox. Eng. Lett. **15**(1), 14 (2007)
37. J.R. Castro, O. Castillo, P. Melin, L.G. Martinez, S. Escobar, I. Camacho, Building fuzzy inference systems with the interval type-2 fuzzy logic toolbox. Adv. Soft Comput. **41**, 53–62 (2007)

38. R. Sepulveda, O. Montiel, G. Lizarraga, O. Castillo, Modeling and simulation of the defuzzification stage of a type-2 fuzzy controller using the Xilinx system generator and Simulink. Stud. Comput. Intell. **257**, 309–325 (2009)
39. B. Widrow, J.R. Glover, Adaptive noise cancelling: principles and applications. IEEE Proc. **63**, 1692–1716 (1975)

Chapter 2
Type-2 Fuzzy Logic Systems

In this chapter, a brief overview of the basic concepts of type-2 fuzzy systems is presented. This overview is intended to provide the basic concepts needed to understand the methods and algorithms presented later in this book [1–3]. The basic concepts that are covered in this chapter are: type-2 fuzzy sets, membership functions, type-2 inference, type reduction and defuzzification.

We begin by defining type-2 fuzzy sets and their corresponding membership functions. If for a type-1 membership function, as in Fig. 2.1, we blur it to the left and to the right, as illustrated in Fig. 2.2, then a type-2 membership function is produced. In this case, for a specific value x', the membership function (u'), takes on different values, which are not all weighted the same, so we can assign membership grades to all of those points.

By doing this for all $x \in X$, we form a three-dimensional membership function—a type-2 membership function—that characterizes a type-2 fuzzy set [2, 3]. A type-2 fuzzy set $\widetilde{A}$, is characterized by the membership function:

$$\tilde{A} = \left\{ \left((x, u), \mu_{\tilde{A}}(x, u) \right) | \forall x \in X, \forall u \in J_x \subseteq [0, 1] \right\} \tag{2.1}$$

in which $0 \leq \mu_{\tilde{A}}(x, u) \leq 1$. In fact $J_x \subseteq [0, 1]$ represents the primary membership of x, and $\mu_{\tilde{A}}(x, u)$is a type-1 fuzzy set known as the secondary set. Hence, a type-2 membership grade can be any subset in [0,1], the primary membership, and corresponding to each primary membership, there is a secondary membership (which can also be in [0,1]) that defines the possibilities for the primary membership. Uncertainty is represented by a region, which is called the footprint of uncertainty (FOU). When $\mu_{\tilde{A}}(x, u) = 1, \forall u \in J_x \subseteq [0, 1]$ we have an interval type-2 membership function, as shown in Fig. 2.3. The uniform shading for the FOU represents the entire interval type-2 fuzzy set and it can be described in terms of an upper membership function$\bar{\mu}_{\tilde{A}}(x)$and a lower membership function $\underline{\mu}_{\tilde{A}}(x)$.

A fuzzy logic system (FLS) described using at least one type-2 fuzzy set is called a type-2 FLS. Type-1 FLSs are unable to directly handle rule uncertainties,

O. Castillo and P. Melin, *Recent Advances in Interval Type-2 Fuzzy Systems*,
SpringerBriefs in Computational Intelligence, DOI: 10.1007/978-3-642-28956-9_2,

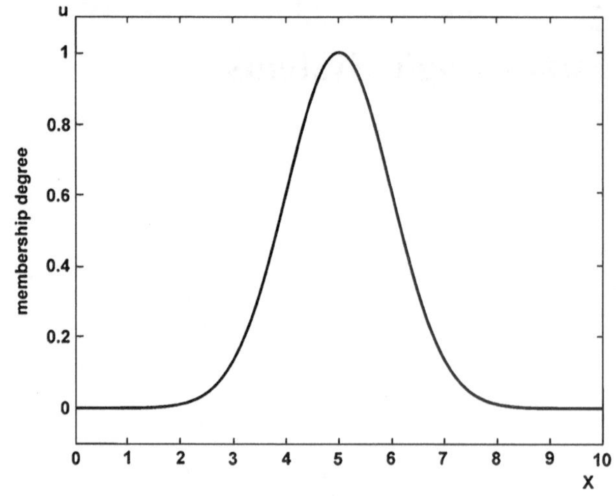

Fig. 2.1 An example of a type-1 membership function

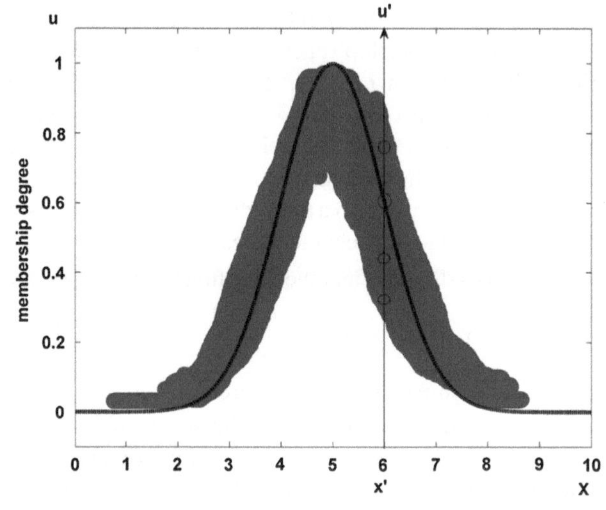

Fig. 2.2 Blurred type-1 membership function

because they use type-1 fuzzy sets that are certain (viz, fully described by single numeric values). On the other hand, type-2 FLSs, are useful in circumstances where it is difficult to determine an exact numeric membership function, and there are measurement uncertainties [3].

A type-2 FLS is characterized by IF–THEN rules, where their antecedent or consequent sets are now of type-2. Type-2 FLSs, can be used when the circumstances are too uncertain to determine exact membership grades such as when the training

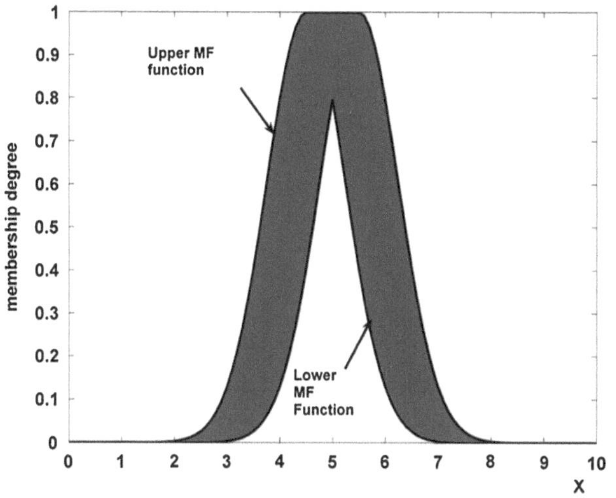

Fig. 2.3 Interval type-2 membership function

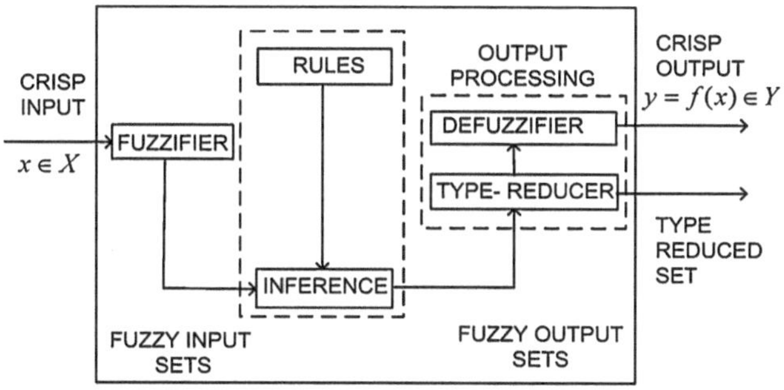

Fig. 2.4 Type-2 fuzzy logic system

data is affected by noise. Similarly, to the type-1 FLS, a type-2 FLS includes a fuzzifier, a rule base, fuzzy inference engine, and an output processor, as we can see in Fig. 2.4 for a Mamdani model. The output processor includes type-reducer and defuzzifier; it generates a type-1 fuzzy set output (from the type-reducer) or a number (from the defuzzifier) [2]. Now we explain each of the blocks shown in Fig. 2.4.

2.1 Fuzzifier

The fuzzifier maps a numeric vector $\mathbf{x} = (x_1,\ldots,x_p)^T \in X_1 x X_2 x \ldots x X_p \equiv \mathbf{X}$ into a type-2 fuzzy set $\tilde{A}_x$ in $\mathbf{X}$ [3], an interval type-2 fuzzy set in this case. We use type-2 singleton fuzzifier, in a singleton fuzzification, the input fuzzy set has only a single point on nonzero membership. $\tilde{A}_x$ is a type-2 fuzzy singleton if $\mu_{\tilde{A}_x}(x) = 1/1$ for $\mathbf{x} = \mathbf{x}'$ and $\mu_{\tilde{A}_x}(x) = 1/0$ for all other $\mathbf{x} \neq \mathbf{x}'$.

2.2 Rules

The structure of rules in a type-1 FLS and a type-2 FLS is the same, but in the latter the antecedents and the consequents is represented by type-2 fuzzy sets. So for a type-2 FLS with p inputs (linguistic variables) $x_1 \in X_1,\ldots,x_p \in X_p$ and one output $y \in Y$, Multiple Input Single Output (MISO), if we assume there are M rules, the lth rule in the type-2 FLS can be written down as follows (where the F's and G are appropriate fuzzy sets for each rule):

$$R^l : \text{IF } x_1 \text{ is } \tilde{F}_1^l \text{and} \cdots \text{and } x_p \text{ is } \tilde{F}_p^l, \text{THEN } y \text{ is } \tilde{G}^l \quad l = 1,\ldots,M \tag{2.2}$$

2.3 Inference

In the type-2 FLS, the inference engine combines rules and gives a mapping from input type-2 fuzzy sets to output type-2 fuzzy sets. It is necessary to compute the join ⊔, (unions) and the meet Π (intersections), as well as the extended sup-star compositions (sup star compositions) of type-2 relations. If $\tilde{F}_1^l \times \cdots \times \tilde{F}_p^l = \tilde{A}^l$, then (2.2) can be re-written as follows

$$R^l : \tilde{F}_1^l \times \cdots \times \tilde{F}_p^l \rightarrow \tilde{G}^l = \tilde{A}^l \rightarrow \tilde{G}^l \quad l = 1,\ldots,M \tag{2.3}$$

R^l is described by the membership function $\mu_{R^l}(\mathbf{x}, y) = \mu_{R^l}(x_1,\ldots,x_p, y)$, where

$$\mu_{R^l}(\mathbf{x}, y) = \mu_{\tilde{A}^l \rightarrow \tilde{G}^l}(\mathbf{x}, y) \tag{2.4}$$

can be written as:

$$\begin{aligned} \mu_{R^l}(\mathbf{x}, y) = \mu_{\tilde{A}^l \rightarrow \tilde{G}^l}(\mathbf{x}, y) &= \mu_{\tilde{F}_1^l}(x_1) \Pi \cdots \Pi \mu_{\tilde{F}_p^l}(x_p) \Pi \mu_{\tilde{G}^l}(y) \\ &= [\Pi_{i=1}^p \mu_{\tilde{F}_i^l}(x_i)] \Pi \mu_{\tilde{G}^l}(y) \end{aligned} \tag{2.5}$$

In general, the p-dimensional input to R^l is given by the type-2 fuzzy set $\tilde{A}_x$whose membership function becomes

$$\mu_{\tilde{A}_x}(\mathbf{x}) = \mu_{\tilde{x}_1}(x_1) \Pi \cdots \Pi \mu_{\tilde{x}p}(x_p) = \Pi_{i=1}^p \mu_{\tilde{x}i}(x_i) \tag{2.6}$$

where $\tilde{X}_i (i = 1,\ldots,p)$ are the labels of the fuzzy sets describing the inputs. Each rule R^l determines a type-2 fuzzy set $\tilde{B}^l = \tilde{A}_x \circ R^l$ such that:

$$\mu_{\tilde{B}^l}(y) = \mu_{\tilde{A}_x \circ R^l} = \sqcup_{x \in X}\left[\mu_{\tilde{A}_x}(\mathbf{x}) \sqcap \mu_{R^l}(\mathbf{x}, y)\right] \quad y \in Y \quad l = 1, \ldots, M \tag{2.7}$$

This dependency is the input/output relation shown in Fig. 2.3, which holds between the type-2 fuzzy set that activates a certain rule in the inference engine and the type-2 fuzzy set at the output of that engine [3].

In the FLS, we used interval type-2 fuzzy sets and intersection under product t-norm, so the result of the input and antecedent operations, which are contained in the firing set $\sqcap_{i=1}^{p}\mu_{\tilde{F}_{i_i}}(x_i' \equiv F^l(\mathbf{x}')$, is an interval type-1 set,

$$F^l(\mathbf{x}') = \left[\underline{f}^l(\mathbf{x}'), \overline{f}^l(\mathbf{x}')\right] \equiv \left[\underline{f}^l, \overline{f}^l\right] \tag{2.8}$$

where

$$\underline{f}^l(\mathbf{x}') = \underline{\mu}_{\tilde{F}_1^l}(x_1') * \cdots * \underline{\mu}_{\tilde{F}_p^l}(x_p') \tag{2.9}$$

and

$$\overline{f}^l(\mathbf{x}') = \overline{\mu}_{\tilde{F}_1^l}(x_1') * \cdots * \overline{\mu}_{\tilde{F}_p^l}(x_p') \tag{2.10}$$

here * stands for the product operation.

2.4 Type Reducer

The type-reducer generates a type-1 fuzzy set output, which is then converted in a numeric output through running the defuzzifier. This type-1 fuzzy set is also an interval set, for the case of our FLS we used center of sets (cos) type reduction, $Y_{\cos}$, which is expressed as [3]

$$\begin{aligned} Y_{\cos}(x) &= [y_l, y_r] \\ &= \int_{y^1 \in [y_l^1, y_r^1]} \cdots \int_{y^M \in [y_l^M, y_r^M]} \int_{f^1 \in [\underline{f}^1, \overline{f}^1]} \cdots \int_{f^M \in [\underline{f}^M, \overline{f}^M]} 1 \Big/ \frac{\sum_{i=1}^{M} f^i y^i}{\sum_{i=1}^{M} f^i} \end{aligned} \tag{2.11}$$

This interval set is determined by its two end points, y_l and y_r, which corresponds to the centroid of the type-2 interval consequent set $\tilde{G}^i$,

$$C_{\tilde{G}^i} = \int_{\theta_1 \in J_{y1}} \cdots \int_{\theta_N \in J_{yN}} 1 \Big/ \frac{\sum_{i=1}^{N} y_i \theta_i}{\sum_{i=1}^{N} \theta_i} = [y_l^i, y_r^i] \tag{2.12}$$

before the computation of $Y_{\cos}$ ($\mathbf{x}$), we must evaluate equation (2.12), and its two end points, y_l and y_r. If the values of f_i and y_i that are associated with y_l are denoted f_l^i and y_l^i, respectively, and the values of f_i and y_i that are associated with y_r are denoted f_r^i and y_r^i, respectively, from equation (2.13), we have [3]

$$y_l = \frac{\sum_{i=1}^{M} f_l^i y_l^i}{\sum_{i=1}^{M} f_l^i} \tag{2.13}$$

$$y_r = \frac{\sum_{i=1}^{M} f_r^i y_r^i}{\sum_{i=1}^{M} f_r^i} \tag{2.14}$$

The values of y_l and y_r define the output interval of the type-2 fuzzy system, which can be used to verify if training or testing data are contained in the output of the fuzzy system. This measure of covering the data is considered as one of the design criteria in finding an optimal interval type-2 FS. The other optimization criteria, is that the length of this output interval should be as small as possible.

2.5 Defuzzifier

From the type-reducer, we obtain an interval set $Y_{\cos}$, to defuzzify it we use the average of y_l and y_r, so the defuzzified output of an interval singleton type-2 FLS is [3]

$$y(\mathbf{x}) = \frac{y_l + y_r}{2} \tag{2.15}$$

To the moment, most of the interval type-2 fuzzy systems that have been developed for the applications follow the architecture of Fig. 2.4 and the definitions presented in this Chapter. In this sense, what has been presented constitutes a good basis for understanding the rest of the chapters of this book.

References

1. O. Castillo, P. Melin, *Soft Computing and Fractal Theory for Intelligent Manufacturing* (Springer, Heidelberg, 2003)
2. O. Castillo, P. Melin, *Type-2 Fuzzy Logic: Theory and Applications* (Springer, Heidelberg, 2008)
3. N.N. Karnik, J.M. Mendel, An introduction to type-2 fuzzy logic systems, Technical Report, University of Southern California (1998)

Chapter 3
Bio-Inspired Optimization Methods

In this chapter a brief overview of the basic concepts from bio-inspired optimization methods needed for this work is presented. In particular, the methods that are covered in this chapter are: particle swarm optimization, genetic algorithms and ant colony optimization.

3.1 Particle Swarm Optimization

Particle swarm optimization is a population based stochastic optimization technique developed by Eberhart and Kennedy in 1995, inspired by social behavior of bird flocking or fish schooling [1]. PSO shares many similarities with evolutionary computation techniques such as the GA [2].

The system is initialized with a population of random solutions and searches for optima by updating generations. However, unlike the GA, the PSO has no evolution operators such as crossover and mutation. In the PSO, the potential solutions, called particles, fly through the problem space by following the current optimum particles [1]. Each particle keeps track of its coordinates in the problem space, which are associated with the best solution (fitness) it has achieved so far (The fitness value is also stored). This value is called *pbest*. Another "best" value that is tracked by the particle swarm optimizer is the best value, obtained so far by any particle in the neighbors of the particle. This location is called *lbest*. When a particle takes all the population as its topological neighbors, the best value is a global best and is called *gbest* [3].

The particle swarm optimization concept consists of, at each time step, changing the velocity of (accelerating) each particle toward its *pbest* and *lbest* locations (local version of PSO). Acceleration is weighted by a random term, with separate random numbers being generated for acceleration toward *pbest* and *lbest* locations [4, 5]. In the past several years, PSO has been successfully applied in

O. Castillo and P. Melin, *Recent Advances in Interval Type-2 Fuzzy Systems*,
SpringerBriefs in Computational Intelligence, DOI: 10.1007/978-3-642-28956-9_3,

many research and application areas. It is demonstrated that PSO gets better results in a faster, cheaper way when compared with other methods [1, 3, 6]. Another reason that PSO is attractive is that there are few parameters to adjust. One version, with slight variations, works well in a wide variety of applications. Particle swarm optimization has been considered for approaches that can be used across a wide range of applications, as well as for specific applications focused on a specific requirement [7–9].

The basic algorithm of PSO has the following nomenclature:

x_z^i— Particle position
v_z^i—Particle velocity
w_{ij}—Inertia weight
p_z^i—Best "remembered" individual particle position
p_z^g—Best "remembered" swarm position
c_1, c_2—Cognitive and Social parameters
r_1, r_2—Random numbers between 0 and 1

The equation to calculate the velocity is:

$$v_{z+1}^i = w_{ij}\, v_z^i + c_1\, r_1\left(p_z^i - x_z^i\right) + c_2\, r_2\left(p_z^g - x_z^i\right) \tag{3.1}$$

and the position of the individual particles is updated as follows:

$$x_{z+1}^i = x_z^i + v_{z+1}^i \tag{3.2}$$

The basic PSO algorithm is defined as follows:

1) *Initialize*

   a) *Set constants* $z_{\max}, c_1, c_2$
   b) *Randomly initialize particle position* $x_0^i \in D$ *in* R^n *for* $i = 1, \ldots, p$
   c) *Randomly initialize particle velocities* $0 \le v_0^i \le v_0^{\max}$ *for* $i = 1, \ldots, p$
   d) *Set* $Z = 1$

2) *Optimize*

   a) *Evaluate function value* f_k^i *using design space coordinates* x_k^i
   b) *If* $f_z^i \le f_{best}^i$ *then* $f_{best}^i = f_z^i, p_z^i = x_z^i$
   c) *If* $f_z^i \le f_{best}^g$ *then* $f_{best}^g = f_z^i, p_z^g = x_z^i$
   d) *If stopping condition is satisfied then go to 3*
   e) *Update all particle velocities* v_z^i for $i = 1, \ldots, p$
   f) *Update al particle positions* x_z^i *for* $i = 1, \ldots, p$
   g) *Increment* z
   (h) *Goto 2(a)*

(3) *Terminate*

3.2 Genetic Algorithms

Genetic Algorithms (GAs) are adaptive heuristic search algorithms based on the evolutionary ideas of natural selection and genetic processes [10]. The basic principles of GAs were first proposed by John Holland in 1975, inspired by the mechanism of natural selection, where stronger individuals are likely the winners in a competing environment [11–13]. GA assumes that the potential solution of any problem is an individual and can be represented by a set of parameters. These parameters are regarded as the genes of a chromosome and can be structured by a string of values in binary form. A positive value, generally known as a fitness value, is used to reflect the degree of "goodness" of the chromosome for the problem, which would be highly related with its objective value. The pseudocode of a GA is as follows:

1. Start with a randomly generated population of n chromosomes (candidate solutions to a problem).
2. Calculate the fitness of each chromosome in the population.
3. Repeat the following steps until n offspring have been created:
   a. Select a pair of parent chromosomes from the current population, the probability of selection being an increasing function of fitness. Selection is done with replacement, meaning that the same chromosome can be selected more than once to become a parent.
   b. With probability (crossover rate), perform crossover to the pair at a randomly chosen point to a form two offspring.
   c. Mutate the two offspring at each locus with probability (mutation rate), and place the resulting chromosomes in the new population.
4. Replace the current population with the new population.
5. Go to step 2.

The simple procedure just described above is the basis for most applications of GAs found in the literature.

3.3 Ant Colony Optimization

Ant Colony Optimization (ACO) is a probabilistic technique that can be used for solving problems that can be reduced to finding good paths along graphs. This method is inspired on the behavior presented by ants in finding paths from the nest or colony to the food source.

The S-ACO is an algorithmic implementation that adapts the behavior of real ants to solutions of minimum cost path problems on graphs [14]. A number of artificial ants build solutions for a certain optimization problem and exchange information about the quality of these solutions making allusion to the communication system of real ants [15].

Let us define the graph $G = (V, E)$, where V is the set of nodes and E is the matrix of the links between nodes. G has $n_G = |V|$ nodes. Let us define L^K as the number of hops in the path built by the ant k from the origin node to the destiny node. Therefore, it is necessary to find:

$$Q = \{q_a, \ldots, q_f | q_1 \in C\} \tag{3.3}$$

where Q is the set of nodes representing a continuous path with no obstacles; $q_a, \ldots, q_f$ are former nodes of the path and C is the set of possible configurations of the free space. If $x^k(t)$ denotes a Q solution in time t, $f(x^k(t))$ expresses the quality of the solution. The S-ACO algorithm is based on Eqs. (3.4–3.6):

$$p_{ij}^k(t) = \begin{cases} \frac{\tau_{ij}^k}{\sum_{j \in N_{ij}^k} \tau_{ij}^\alpha(t)} & if\ j \in N_i^k \\ 0 & if\ j \notin N_i^k \end{cases} \tag{3.4}$$

$$\tau_{ij}(t) \leftarrow (1 - \rho)\tau_{ij}(t) \tag{3.5}$$

$$\tau_{ij}(t+1) = \tau_{ij}(t) + \sum_{k=1}^{n_k} \tau_{ij}(t) \tag{3.6}$$

Equation (3.4) represents the probability for an ant k located on a node i selects the next node denoted by j, where, N_i^k is the set of feasible nodes (in a neighborhood) connected to node i with respect to ant k, $\tau_{ij}$is the total pheromone concentration of link ij, and α is a positive constant used as a gain for the pheromone influence.

Equation (3.5) represents the evaporation pheromone update, where $\rho \in [0, 1]$ is the evaporation rate value of the pheromone trail. The evaporation is added to the algorithm in order to force the exploration of the ants, and avoid premature convergence to sub-optimal solutions [16]. For $\rho = 1$ the search becomes completely random [17].

Equation (3.6), represents the concentration pheromone update, where $\Delta\tau_{ij}^k$ is the amount of pheromone that an ant k deposits in a link ij in a time t.

The general steps of S-ACO are the following:

1. *Set a pheromone concentration* τ_{ij} *to each link (i,j).*
2. *Place a number* $k = 1, 2, \ldots, n$ *in the nest.*
3. *Iteratively build a path to the food source (destiny node), using Eq. (3.4) for every ant.*

- *Remove cycles and compute each route weight* $f(x^k(t))$. *A cycle could be generated when there are no feasible candidates nodes, that is, for any i and any k,* $N_i^k = \emptyset$,*then the predecessor of that node is included as a former node of the path.*

4. *Apply evaporation using Eq. (3.5).*
5. *Update of the pheromone concentration using Eq. (3.6)*
6. *Finally, finish the algorithm in any of the three different ways:*
   - *When a maximum number of epochs has been reached.*
   - *When it has found an acceptable solution, with $f(x_k(t)) < \varepsilon$.*
   - *When all ants follow the same path.*

3.4 General Remarks About Optimization of Type-2 Fuzzy Systems Using Bio-Inspired Methods

The problem of designing type-2 fuzzy systems can be solved with any of the above mentioned optimization methods. The main issue in any of these methods is deciding on the representation of the type-2 fuzzy system in the corresponding optimization paradigm. For example, in the case of GAs, the type-2 fuzzy systems must be represented in the chromosomes. On the other hand, in PSO the fuzzy system is represented as a particle in the optimization process. In the ACO method, the fuzzy system can be represented as one of the paths that the ants can follow in a graph. Also, the evaluation of the fuzzy system must be represented as an objective function in any of the methods. In this paper, we offer a comprehensive review of the most representative works in optimization of type-2 fuzzy systems that have been done around the world.

References

1. R. Martinez, A. Rodriguez, O. Castillo, L.T. Aguilar, Type-2 fuzzy logic controllers optimization using genetic algorithms and particle swarm optimization, in *Proceedings of the IEEE International Conference on Granular Computing, GrC*, 2010, pp. 724–727
2. K.-J. Park, S.-K. Oh, W. Pedrycz, Design of interval type-2 fuzzy neural networks and their optimization using real-coded genetic algorithms, in *Proceedings of the IEEE Conference on Fuzzy Systems*, Jeju, Korea, 2009, pp. 2013–2018
3. R. Martinez, O. Castillo, L.T. Aguilar, A. Rodriguez, Optimization of type-2 fuzzy logic controllers using PSO applied to linear plants. Stud. Comput. Intell. **318**, 181–193 (2010)
4. R.A. Aliev, W. Pedrycz, B.G. Guirimov, R.R. Aliev, U. Ilhan, M. Babagil, S. Mammadli, Type-2 fuzzy neural networks with fuzzy clustering and differential evolution optimization. Inf. Sci. **181**(9), 1591–1608 (2011)
5. M.A. Khanesar, M. Teshnehlab, E. Kayacan, O. Kaynak, A novel type-2 fuzzy membership function: Application to the prediction of noisy data, in *Proceedings of the IEEE International Conference on Computational Intelligence for Measurement Systems and Applications, CIMSA 2010*, 2010, pp. 128–133
6. W.-H.R. Jeng, C.-Y. Yeh, S.-J. Lee, General type-2 fuzzy neural network with hybrid learning for function approximation, in *Proceedings of the IEEE Conference on Fuzzy Systems*, Jeju, Korea, 2009, pp. 1534–1539

7. J. Cao, P. Li, H. Liu, D. Brown, Adaptive fuzzy controller for vehicle active suspensions with particle swarm optimization, in *Proceedings of SPIE—The International Society of Optical Engineering*, 2008, p. 7129
8. G.-S. Kim, I.-S. Ahn, S.-K. Oh, The design of optimized fuzzy neural networks and its application. Trans. Korean Inst. Electr. Eng. **58**(6), 1615–1623 (2009)
9. X.-Z. Zhao, Y.-B. Gao, J.-F. Zeng, Y.-P. Yang, PSO type-reduction method for geometric interval type-2 fuzzy logic systems. J. Harbin Inst. Technol. **15**(6), 862–867 (2008)
10. O. Cordon, F. Gomide, F. Herrera, F. Hoffmann, L. Magdalena, Ten years of genetic fuzzy systems: current framework and new trends. Fuzzy Sets Syst. **141**, 5–31 (2004)
11. O. Castillo, P. Melin, *Soft Computing for Control of Non-Linear Dynamical Systems* (Springer, Heidelberg, 2001)
12. T.W. Chua, W.W. Tan, Genetically evolved fuzzy rule-based classifiers and application to automotive classification. Lecture Notes in Computer Science, vol. 5361 (2008), pp. 101–110
13. O. Cordon, F. Herrera, P. Villar, Analysis and guidelines to obtain a good uniform fuzzy partition granularity for fuzzy rule-based systems using simulated annealing. Int. J. Approx. Reason. **25**, 187–215 (2000)
14. C.-F. Juang, C.-H. Hsu, Reinforcement interval type-2 fuzzy controller design by online rule generation and Q-value-aided ant colony optimization. IEEE Trans. Syst. Man Cybern. B Cybern. **39**(6), 1528–1542 (2009)
15. O. Castillo, R. Martinez-Marroquin, P. Melin, F. Valdez, J. Soria, Comparative study of bio-inspired algorithms applied to the optimization of type-1 and type-2 fuzzy controllers for an autonomous mobile robot. Info. Sci. **192**(1), 19–38 (2012)
16. C.-F. Juang, C.-H. Hsu, C.-F. Chuang, Reinforcement self-organizing interval type-2 fuzzy system with ant colony optimization, in *Proceedings of IEEE International Conference on Systems, Man and Cybernetics*, San Antonio, 2009, pp. 771–776
17. R. Martinez-Marroquin, O. Castillo, J. Soria, Parameter tuning of membership functions of a type-1 and type-2 fuzzy logic controller for an autonomous wheeled mobile robot using ant colony optimization, in *Proceedings of IEEE International Conference on Systems, Man and Cybernetics*, San Antonio, 2009, pp. 4770–4775

Chapter 4
Overview of Genetic Algorithms Applied in the Optimization of Type-2 Fuzzy Systems

There have been many works reported in the literature optimizing type-2 fuzzy systems using different kinds of genetic algorithms. Most of these works have had relative success according to the different areas of application. In this chapter, we offer a representative review of these types of works to illustrate the advantages of using a bio-inspired optimization technique for automating the design process of type-2 fuzzy systems. This overview has the goal of providing the reader with an idea of the diversity of applications that have been achieved using genetic algorithms for type-2 fuzzy system optimization.

In a paper by Park et al. [1] a design methodology of interval type-2 fuzzy neural networks (IT2FNN) was introduced to optimize the network using a real-coded genetic algorithm. IT2FNN is the combination between the fuzzy neural network (FNN) and interval type-2 fuzzy set with uncertainty. The antecedent part of the network is composed of the fuzzy division of input space and the consequence part of the network is represented by polynomial functions. The parameters such as the apexes of membership function, uncertainty parameter, the learning rate and the momentum coefficient are optimized using a Genetic Algorithm (GA). The proposed network is evaluated with the performance between the approximation and the generalization abilities.

In a work by Chua and Tan [2] a method for genetically evolving type-2 fuzzy rule based classifiers was proposed. This work was aimed at investigating if type-2 fuzzy classifiers can deliver a better performance when there exists an imprecise decision boundary caused by improper feature extraction method. A GA is used to tune the fuzzy classifiers under Pittsburgh scheme. The proposed fuzzy classifiers were successfully applied to an automotive application whereby the classifier needs to detect the presence of human in a vehicle. Results revealed that a type-2 classifier has the edge over type-1 classifier when the decision boundaries are imprecise and the fuzzy classifier itself has not enough degrees of freedom to construct a suitable boundary. Conversely, when decision boundaries are clear, the advantage of type-2 framework may not be significant anymore. In any case, the

O. Castillo and P. Melin, *Recent Advances in Interval Type-2 Fuzzy Systems*,
SpringerBriefs in Computational Intelligence, DOI: 10.1007/978-3-642-28956-9_4,

performance of a type-2 fuzzy classifier is at least comparable with a type-1 fuzzy classifier. When dealing with real world classification problem where the uncertainty is usually difficult to be estimated, type-2 fuzzy classifier can be a more rational choice.

In a paper by Cazarez et al. [3] a genetic-type-2 fuzzy approach was proposed to optimize the parameters of the Membership Functions (MFs) of a Type-2 Fuzzy Logic System (FLS) applied to control. The chromosome was designed to represent the parameters of the MFs of a pre-established Type-2 FLS. A case of study was proposed to evaluate the optimization process, which was to achieve the output regulation problem of a servomechanism with backlash. The problem is the design of a type-2 fuzzy logic controller which was optimized by a GA to obtain the closed-loop system in which the load of the driver is regulated to a desired position. Simulations results illustrate the effectiveness of the optimized closed-loop system.

In the work of Lopez et al. [4] a new method for response integration in ensemble neural networks with type-2 fuzzy logic using genetic algorithms for optimization was proposed. In this paper, pattern recognition with ensemble neural networks for the case of fingerprints was considered. An ensemble neural network of three modules was used. Each module was a local expert on person recognition based on its biometric measure (pattern recognition for fingerprints). The response integration method of the ensemble neural networks has the goal of combining the responses of the modules to improve the recognition rate of the individual modules. Using GAs to optimize the membership functions the results of the type-2 fuzzy systems were improved. In this paper the results of a type-2 approach for response integration were shown to outperform the type-1 logic approach.

In the work of Cai et al. [5] a novel fuzzy-neural network combining a Type-2 Fuzzy Logic System (FLS) and a Genetic Algorithm (GA) based on a Takagi–Sugeno–Kang fuzzy neural network (GA-TSKfnn), is presented. The rational for this combination is that type-2 fuzzy sets are better able to deal with rule uncertainties, while the optimal GA-based tuning of the T2GA-TSKfnn parameters achieves better classification results. However, a general T2GA-TSKfnn is computationally very intensive due to the complexity of the type-2 to type-1 reduction. Therefore, an interval T2GA-TSKfnn implementation to simplify the computational process was adopted. Simulation results were provided to compare the T2GA-TSKfnn against other fuzzy neural networks. These results show that the proposed system is able to achieve a higher classification rate when compared against a number of other traditional neuro-fuzzy classifiers.

In the work of Wagner and Hagras, [6, 7] a genetic algorithm for evolving type-2 fuzzy logic controllers for real world autonomous robots was presented. The type-2 Fuzzy Logic Controller (FLC) has started to emerge as a promising control mechanism for autonomous mobile robots navigating in real world environments. This is because such robots need control mechanisms such as type-2 FLCs which can handle the large amounts of uncertainties present in real world environments. However, manually designing and tuning the type-2 Membership

Functions (MFs) for an interval type-2 FLC to give a good response is a difficult task. This work describes a genetic algorithm to evolve the type-2 MFs of interval type-2 FLCs for mobile robots that will navigate in real world environments. The GA based system converges after a small number of iterations to type-2 MFs which give a very good performance. A series of real world experiments in which the evolved type-2 FLCs controlled a real robot in an outdoor arena was performed. The evolved type-2 FLCs dealt with the uncertainties present in the real world to give a very good performance that has outperformed their type-1 counterparts as well as the manually designed type-2 FLCs.

In the work of Qiu et al. [8] statistical genetic interval valued fuzzy systems for prediction in clinical trials are presented. In recent years, statistical tools and computational intelligence methods have played important roles in many areas. After statistically optimizing interval-valued fuzzy membership functions in the type-2 fuzzy logic system, genetic algorithms were applied to optimize them. The proposed method was used to predict survival times for patients in clinical trials. The results show that the new GA-based method was more accurate than traditional type-1 and type-2 methods.

In the work by Tan and Wu [9] the design of type reduction strategies for type-2 fuzzy logic systems using genetic algorithms was presented. While a type-2 fuzzy system has the capability to model more complex relationships, the output of a type-2 fuzzy inference engine is a type-2 fuzzy set that needs to be type-reduced before defuzzification can be performed. Unfortunately, type-reduction is usually achieved using the computationally intensive Karnik–Mendel iterative algorithm. In order for type-2 fuzzy systems to be useful for real-time applications, the computational burden of type-reduction needs to be relieved. This work was aimed at designing computationally efficient type-reducers using a genetic algorithm. The proposed type-reducer is based on the concept known as equivalent type-1 fuzzy systems (ET1FSs), a collection of type-1 FSs that replicates the input–output relationship of a type-2 fuzzy system. By replacing a type-2 fuzzy system with a collection of ET1FSs, the type-reduction process then simplifies to deciding which ET1FS to employ in a particular situation. The strategy for selecting the ET1FS is evolved by a GA. Results were presented to demonstrate that the proposed type-reducing algorithm has lower computational cost and may provide better performance than FLSs that employ existing type-reducers.

In the work by Wu and Tan [10] genetic learning and performance evaluation of interval type-2 fuzzy logic controllers was presented. Type-2 fuzzy sets, which are characterized by membership functions that are themselves fuzzy, have been attracting interest. This paper focuses on advancing the understanding of interval (FLCs). First, a type-2 FLC was evolved using genetic algorithms. The type-2 FLC was then compared with another three GA evolved type-1 FLCs that have different design parameters. The objective was to examine the amount by which the extra degrees of freedom, provided by antecedent type-2 fuzzy sets, was able to improve the control performance. Experimental results show that better control can be achieved using a type-2 FLC with fewer fuzzy sets/rules so one benefit of type-2 FLC was a lower trade-off between modeling accuracy and interpretability.

The work by Wu and Tan [11] focuses on evolving type-2 fuzzy logic controllers genetically and examining whether they are better able to handle modeling uncertainties. The study was conducted by utilizing a type-2 FLC, evolved by a genetic algorithm, to control a liquid-level process. A two stage strategy is employed to design the type-2 FLC. First, the parameters of a type-1 FLC are optimized using the GA. Next, the footprint of uncertainty was evolved by blurring the fuzzy input set. Experimental results show that the type-2 FLC copes well with the complexity of the plant, and can handle the modeling uncertainty better than its type-1 counterpart.

In the work by Wang et al. [12] a type-2 fuzzy logic system cascaded with neural network, Type-2 Fuzzy Neural Network (T2FNN), was presented to handle uncertainty with dynamical optimal learning. A T2FNN consists of a type-2 fuzzy linguistic process as the antecedent part, and the two-layer interval neural network as the consequent part. A general T2FNN is computational-intensive due to the complexity of type-2 to type-1 reduction. Therefore, the interval T2FNN is adopted in this work to simplify the computational process. The dynamical optimal training algorithm for the two-layer consequent part of interval T2FNN was first developed. The stable and optimal left and right learning rates for the interval neural network, in the sense of maximum error reduction, can be derived for each iteration in the training process (back propagation). It can also be shown that both learning rates cannot be both negative. Further, due to variation of the initial MF parameters, i.e., the spread level of uncertain means or deviations of interval Gaussian MFs, the performance of back propagation training process may be affected. To achieve better total performance, a genetic algorithm was designed to search optimal spread rate for uncertain means and optimal learning for the antecedent part. Several examples are fully illustrated. Excellent results are obtained for the truck backing-up control and the identification of nonlinear system, which yield more improved performance than those using type-1 FNN.

In the work by Innocent et al. [13] the exploratory use of type 2 fuzzy sets to represent the perceptions of lung scan images by experts in order to predict pulmonary emboli using type 2 fuzzy relations is presented. A genetic algorithm was used to find suitable parameters for the fuzzy sets so that a good classification was achieved. Preliminary results with a limited data set demonstrating the potential power of the approach were presented.

In the work by Cervantes and Castillo [14] a genetic design of a fuzzy system for the longitudinal control of an F-14 airplane was presented. The longitudinal control is carried out only by controlling the elevators of the airplane. To carry out such monitoring it is necessary to use the stick, the rate of elevation and the angle of attack. These three variables are the inputs into the fuzzy inference system, which is of Mamdani type, and the output the values of the elevators are obtained. Simulation results of the longitudinal control are obtained using a plant in Simulink and those results were compared against the PID controller. Genetic algorithms were used to optimize parameters of type-2 and type-1 fuzzy systems to find the best fuzzy controller under noisy conditions. The type-2 fuzzy controller outperforms the type-1 when the level of noise is sufficiently high.

In the work by Sanchez and Melin [15] a Modular Neural Network (MNN) for iris, ear and voice recognition was presented. The proposed MNN architecture consists of three modules, one for each biometric measure: iris, ear and voice. Each module is divided into other three sub modules. Each sub module contains different information, which consists of the database divided in three parts. The integration of each biometric measure was considered separately. Later, the integration of the modules was performed with a fuzzy logic integrator. Also, the optimization of the modular neural networks and the fuzzy integrators was performed using genetic algorithms, and comparisons were made between optimized results and the results without optimization. The use of type-2 fuzzy logic was considered in the fuzzy response integrators, and the result was that that higher recognition rates under noisy conditions were achieved with a significant improvement over type-1 fuzzy logic.

In the work of Martinez et al. [16], a tracking controller for the dynamic model of a unicycle mobile robot by integrating a kinematic and a torque controller based on type-2 fuzzy logic theory and genetic algorithms was proposed. Genetic optimization enables finding the optimal parameters of the type-2 fuzzy controller for the mobile robot. Computer simulations are presented confirming the performance of the tracking controller and its application to different navigation problems.

In the work of Hidalgo et al. [17], type-2 fuzzy inference systems as integration methods in modular neural networks for multimodal biometry were proposed. In this work a comparative study between fuzzy inference systems as methods of integration in modular neural networks for multimodal biometry was presented. These methods of integration are based on techniques of type-1 and type-2 fuzzy logic. Also, the fuzzy systems are optimized with simple genetic algorithms with the goal of having optimized versions of both types of fuzzy systems. First, the use of type-1 fuzzy logic and later the approach with type-2 fuzzy logic were considered. The fuzzy systems were developed using genetic algorithms to handle fuzzy inference systems with different membership functions, like the triangular, trapezoidal and Gaussian; since these algorithms can generate fuzzy systems automatically. Then the response integration of the modular neural network was tested with the optimized fuzzy systems of integration. The comparative study of the type-1 and type-2 fuzzy inference systems was made to observe the behavior of the two different integration methods for modular neural networks for multimodal biometry.

In Table 4.1 a summary of the previously presented contributions, where GAs have been applied to optimize type-2 fuzzy systems, is presented. The comparison shown in Table 4.1 is based on the following criteria: author names, year of publication, reference number, domain of the problem, if a comparison with type-1 fuzzy logic is provided, if a comparison with other optimization methods is presented, and why type-2 fuzzy logic was used by the authors. From Table 4.1 it can be noted that most of the applications have been in designing optimal type-2 fuzzy systems (with genetic algorithms) for intelligent control and pattern recognition, with fewer applications in prediction and classification.

Table 4.1 GAs for the optimization of type-2 fuzzy systems

Author(s) (pub. year)	Ref. no.	Domain of the problem	Comparison with type-1	Comparison with other optimization	Why type-2 is required for the problem?
Park et al. 2009	[1]	Theory	No	No	Improving approximation
Chua and Tan 2008	[2]	Classification	Yes	No	Uncertainty in classification
Cazarez et al. 2008	[3]	Control	Yes	No	Uncertainty in control
Lopez et al. 2008	[4]	Pattern recognition	Yes	No	Uncertainty in recognition
Cai et al. 2007	[5]	Classification	Yes	Yes	Uncertainty in classification
Wagner and Hagras 2007	[6,7]	Control	Yes	No	Uncertainty in control
Qiu et al. 2007	[8]	Prediction	Yes	No	Uncertainty in prediction
Tan and Wu 2007	[9]	Theory	Yes	No	Improving type reduction
Wu and Tan 2006	[10]	Control	Yes	No	Testing type-2 fuzzy control
Wu and Tan 2004	[11]	Control	Yes	No	Uncertainty in control
Wang et al. 2004	[12]	Control	Yes	Yes	Uncertainty in control
Innocent et al. 2001	[13]	Prediction	Yes	No	Uncertainty in prediction
Cervantes and Castillo 2010	[14]	Control	Yes	No	Uncertainty in control
Sanchez and Melin 2010	[15]	Pattern recognition	Yes	No	Uncertainty in recognition
Martinez et al. 2009	[16]	Control	Yes	Yes	Uncertainty in robot control
Hidalgo et al. 2009	[17]	Pattern recognition	Yes	No	Uncertainty in recognition

To the moment, genetic algorithms have been the most used optimization method for obtaining optimal designs of type-2 fuzzy systems. However, more recently other methods, like PSO and ACO have been also used with success. In the following chapters the use of PSO and ACO in optimizing type-2 fuzzy systems will be reviewed in detail

References

1. K.-J. Park, S.-K. Oh, W. Pedrycz, Design of interval type-2 fuzzy neural networks and their optimization using real-coded genetic algorithms, in *Proceedings of the IEEE Conference on Fuzzy Systems*, Jeju, Korea, 2009, pp. 2013–2018
2. T.W. Chua, W.W. Tan, Genetically evolved fuzzy rule-based classifiers and application to automotive classification. Lecture Notes in Computer Science, vol. 5361 (2008), pp. 101–110
3. N.R. Cazarez-Castro, L.T. Aguilar, O. Castillo, Genetic optimization of a type-2 fuzzy controller for output regulation of a servomechanism with backlash, in *Proceedings of the International Conference on Electrical Engineering, Computing Science and Automatic Control CCE 2008*, Mexico, 2008, pp. 268–273
4. M. Lopez, P. Melin, O. Castillo, Optimization of response integration with fuzzy logic in ensemble neural networks using genetic algorithms. Stud. Comput. Intell. **154**, 129–150 (2008)
5. A. Cai, C. Quek, D.L. Maskell, Type-2 GA-TSK fuzzy neural network, in *Proceedings of IEEE Congress on Evolutionary Computation, CEC 2007*, 2007, pp. 1578–1585
6. C. Wagner, H. Hagras, A genetic algorithm based architecture for evolving type-2 fuzzy logic controllers for real world autonomous mobile robots, in *Proceedings of the IEEE Conference on Fuzzy Systems*, London, 2007
7. C. Wagner, H. Hagras, Evolving type-2 fuzzy logic controllers for autonomous mobile robots. Adv. Soft Comput. **41**, 16–25 (2007)
8. Y. Qiu, Y.-Q. Zhang, Y. Zhao, Statistical genetic interval-valued fuzzy systems with prediction in clinical trials, in *Proceedings of the IEEE International Conference on Granular Computing*, San Jose, 2007, pp. 129–132
9. W.-W. Tan, D. Wu, Design of type-reduction strategies for type-2 fuzzy logic systems using genetic algorithms. Stud. Comput. Intell. **66**, 169–187 (2007)
10. D. Wu, W.-W. Tan, Genetic learning and performance evaluation of interval type-2 fuzzy logic controllers. Eng. Appl. Artif. Intell. **19**(8), 829–841 (2006)
11. D. Wu, W.-W. Tan, A type-2 fuzzy logic controller for the liquid level process, in *Proceedings of the IEEE Conference on Fuzzy Systems*, Budapest, 2004, pp. 953–958
12. C.-H. Wang, C.-S. Cheng, T.-T. Lee, Dynamical optimal training for interval type-2 fuzzy neural network (T2FNN). IEEE Trans. Syst. Man Cybern. B Cybern. **34**(3), 1462–1477 (2004)
13. P.R. Innocent, R.I. John, I. Belton, D. Finlay, Type-2 fuzzy representations of lung scans to predict pulmonary emboli, in *Proceedings of the Annual Conference of the North American Fuzzy Information Processing Society, NAFIPS 2001*, Vancouver, 2001, pp. 1902–1907
14. L. Cervantes, O. Castillo, Design of a fuzzy system for the longitudinal control of an F-14 airplane. Stud. Comput. Intell. **318**, 213–224 (2010)
15. D. Sanchez, P. Melin, Modular neural network with fuzzy integration and its optimization using genetic algorithms for human recognition based on iris, ear and voice biometrics. Stud. Comput. Intell. **312**, 85–102 (2010)
16. R. Martinez, O. Castillo, L.T. Aguilar, Optimization of interval type-2 fuzzy logic controllers for a perturbed autonomous wheeled mobile robot using genetic algorithms. Inf. Sci. **179**(13), 2158–2174 (2009)
17. D. Hidalgo, O. Castillo, P. Melin, Type-1 and type-2 fuzzy inference systems as integration methods in modular neural networks for multimodal biometry and its optimization with genetic algorithms. Inf. Sci. **179**(13), 2123–2145 (2009)

References

[illegible]

Chapter 5
Particle Swarm Optimization in the Design of Type-2 Fuzzy Systems

There have been several works reported in the literature optimizing type-2 fuzzy systems using different kinds of PSO algorithms. Most of these works have had relative success according to the different areas of application. In this chapter, we offer a representative review of these types of works to illustrate the advantages of using the PSO optimization technique for automating the design process of type-2 fuzzy systems.

In the work of Al-Jaafreh and Al-Jumaily [1], a training method for a type-2 fuzzy system using PSO was presented. This work presents the improvement and implementation for two recent intelligent techniques; Type-2 Fuzzy System (T2 FS) and particle swarm optimization and presents a new method to optimize parameters of the primary membership functions of T2 FS using PSO to improve the performance and increase the accuracy of the T2 FS model. The implementation of the suggested method on mean blood pressure estimation has a very successful rate.

In the work of Zhao et al. [2], a PSO type-reduction method for geometric interval type-2 fuzzy logic systems based on the particle swarm optimization algorithm was presented. With the PSO type-reduction, the inference principle of geometric interval FLS operating on the continuous domain is consistent with that of traditional interval type-2 FLS operating on the discrete domain. With comparative experiments, it is proved that the PSO type-reduction exhibits good performance, and is a satisfactory complement for the theory of geometric interval type-2 fuzzy logic systems.

In the work of Cao et al. [3], the PSO algorithm was used to derive an Adaptive Fuzzy Logic Controller (AFC) based on interval fuzzy membership functions for vehicle non-linear active suspension systems. The interval membership functions were utilized in the AFC design to deal with not only non-linearity and uncertainty caused from irregular road inputs and immeasurable disturbance, but also the potential uncertainty of expert's knowledge and experience. The adaptive strategy was designed to self-tune the active force between the lower bounds and upper bounds of interval fuzzy outputs. A case study based on a quarter active

O. Castillo and P. Melin, *Recent Advances in Interval Type-2 Fuzzy Systems*, SpringerBriefs in Computational Intelligence, DOI: 10.1007/978-3-642-28956-9_5,

suspension model demonstrated that the proposed adaptive fuzzy controller significantly outperforms conventional fuzzy controllers of an active suspension and a passive suspension.

In the work of Kim et al. [4], the design of optimized type-2 fuzzy neural networks using PSO was presented. In order to develop reliable on-site Partial Discharge (PD) pattern recognition algorithm, Type-2 Fuzzy Neural Networks (T2FNNs) optimized by means of particle swarm optimization were introduced. T2FNNs exploit type-2 fuzzy sets which have a characteristic of robustness in the diverse area of intelligence systems. Considering the on-site situation where it is not easy to obtain voltage phases to be used for Phase Resolved Partial Discharge Analysis, the PD data sets measured in the laboratory were artificially changed into data sets with shifted voltage phases and added noise in order to test the proposed algorithm. Also, the results obtained by the proposed algorithm were compared with that of conventional neural networks as well as the existing radial basis function neural networks. The T2FNNs proposed in this study appeared to have better performance when compared to conventional neural networks.

In the work by Martinez et al. [5], bio-inspired optimization methods were applied to design type-2 fuzzy logic controllers to minimize the steady state error of linear plants. In particular, the optimal type-2 fuzzy controllers obtained with genetic algorithms and PSO were compared using benchmark plants. The bio-inspired methods were used to find the parameters of the membership functions of the type-2 fuzzy system to obtain the optimal controller. Simulation results were presented to show the feasibility of the proposed approaches. Both GAs and PSO were able to achieve optimal design for the benchmark plants.

In the work of Jeng et al. [6], a novel Takagi–Sugeno–Kang type fuzzy neural network that uses general type-2 fuzzy sets, called General Type-2 Fuzzy Neural Network (GT2FNN), was proposed for function approximation. The problems of constructing a GT2FNN include type reduction, structure identification, and parameter identification. An efficient strategy was proposed by using α-cuts to decompose a general type-2 fuzzy set into several interval type-2 fuzzy sets to solve the type reduction problem. Incremental similarity based fuzzy clustering and linear least squares regression were combined to solve the structure identification problem. Regarding the parameter identification, a hybrid learning algorithm which combines PSO and a recursive least squares estimator was proposed for refining the antecedent and consequent parameters, respectively, of the fuzzy rules.

In the work by Martinez et al. [7], the optimization of type-2 fuzzy logic controllers using PSO was presented. The PSO method was applied to find the parameters of the membership functions of an interval type-2 fuzzy logic controller in order to minimize the steady state error for linear systems. PSO was used to find the optimal interval type-2 fuzzy controller to achieve regulation of the output and stability of the closed-loop system. For this purpose, the values of the cognitive, social and inertia parameter in the PSO algorithm were changed. Simulation results, with the optimal type-2 fuzzy controller implemented in Simulink, show the potential applicability of the proposed approach. The PSO algorithm achieved good results with fast execution times.

In the work by Khanesar et al. [8], a novel, diamond-shaped type-2 fuzzy membership function was introduced. The proposed type-2 fuzzy membership function has certain values on 0 and 1, but it has some uncertainties for the other membership values. It has been shown that the type-2 fuzzy system using this type of membership function has some noise reduction property in the presence of noisy inputs. The appropriate parameter selection to be able to achieve noise reduction property was also considered. A hybrid method consisting of PSO and gradient descent algorithm was used to optimize the parameters of the proposed type-2 fuzzy system. PSO is a derivative-free optimizer, and the possibility of the entrapment of this optimizer in local minimums is less than the gradient descent method. The proposed type-2 fuzzy system and the hybrid parameter estimation method were then tested on the prediction of a noisy, chaotic dynamical system. The simulation results show that the type-2 fuzzy predictor with the proposed novel membership functions shows a superior performance when compared to the other existing type-2 fuzzy systems in the presence of noisy inputs.

In this work of Bingül and Karahan [9], two-degrees of freedom planar robot was controlled by fuzzy logic controller tuned with a particle swarm optimization. For a given trajectory, the parameters of Mamdani-type-Fuzzy Logic Controller (the centers and the widths of the Gaussian membership functions in inputs and output) were optimized by the particle swarm optimization with three different cost functions. In order to compare the optimized fuzzy logic controller with different controllers, the PID controller was also tuned with particle swarm optimization. In order to test the robustness of the tuned controllers, the model parameters and the given trajectory were changed and the white noise was added to the system. The simulation results show that fuzzy logic controller tuned by particle swarm optimization is better and more robust than the PID tuned by particle swarm optimization for robot trajectory control.

In the work by Oh et al. [10], the design methodology of an optimized fuzzy controller with the aid of particle swarm optimization (PSO) for ball and beam system was introduced. The ball and beam system is a well-known control engineering experimental setup which consists of servo motor, beam and ball. This system exhibits a number of interesting and challenging properties when being considered from the control perspective. The ball and beam system determines the position of ball through the control of a servo motor. The displacement change of the position of ball leads to the change of the angle of the beam which determines the position angle of a servo motor. The fixed membership function design of type-1 based fuzzy logic controller (FLC) leads to the difficulty of rule-based control design when representing linguistic nature of knowledge. In type-2 FLC as the expanded type of type-1 FL, we can effectively improve the control characteristic by using the footprint of uncertainty (FOU) of the membership functions. Type-2 FLC exhibits some robustness when compared with type-1 FLC. Through computer simulation as well as real-world experiment, we apply optimized type-2 fuzzy cascade controllers based on PSO to ball and beam system. To evaluate performance of each controller, we consider controller characteristic parameters such as maximum overshoot, delay time, rise time, settling time, and a steady-state

Table 5.1 PSO for type-2 fuzzy system optimization

Author(s) (pub. year)	Ref. no.	Domain of the problem	Comparison with type-1	Comparison with other optimization	Why type-2 is required for the problem?
Al-Jaafreh and Al-Jumaily 2007	[1]	Prediction	No	No	Uncertainty in estimation
Zhao et al. 2008	[2]	Theory	No	No	Improving type reduction
Cao et al. 2008	[3]	Control	Yes	No	Uncertainty in control
Kim et al. 2009	[4]	Pattern recognition	No	Yes	Uncertainty in recognition
Martinez et al. 2010	[5]	Control	Yes	Yes	Uncertainty in control
Jeng et al. 2009	[6]	Theory	No	No	Uncertainty in identification
Martinez et al. 2010	[7]	Control	Yes	Yes	Testing type-2 fuzzy control
Khanesar et al. 2010	[8]	Prediction	Yes	No	Uncertainty in prediction
Bingul and Karahan 2011	[9]	Control	No	No	Presence of noise
Oh et al. 2011	[10]	Control	Yes	Yes	Testing type-2 fuzzy control

error. In the sequel, the optimized fuzzy cascade controller is realized and also experimented with through running two detailed comparative studies including type-1/type-2 fuzzy controller and genetic algorithms/particle swarm optimization.

In Table 5.1 a summary of the previously presented contributions, where PSO has been applied to optimize type-2 fuzzy systems, is presented. The comparison shown in Table 5.1 is based on the following criteria: author names, publication year, reference number, domain of the problem, if a comparison with type-1 fuzzy logic is provided, if a comparison with other optimization methods is presented, and why type-2 fuzzy logic was used by the authors. From Table 5.1 it can be noted that most of the applications of PSO in designing optimal type-2 fuzzy systems have been in the area of intelligent control, with fewer applications in pattern recognition and time series prediction. It can also be noted that the number of papers using PSO is lower than the ones using GAs, mentioned in the previous section.

References

1. M.O. Al-Jaafreh, A.A. Al-Jumaily, Training type-2 fuzzy system by particle swarm optimization, in *IEEE Congress on Evolutionary Computation, CEC 2007*, Singapore, 2007, pp. 3442–3446
2. X.-Z. Zhao, Y.-B. Gao, J.-F. Zeng, Y.-P. Yang, PSO type-reduction method for geometric interval type-2 fuzzy logic systems. J. Harbin Inst. Technol. **15**(6), 862–867 (2008)

3. J. Cao, P. Li, H. Liu, D. Brown, Adaptive fuzzy controller for vehicle active suspensions with particle swarm optimization, in *Proceedings of SPIE—The International Society of Optical Engineering*, 2008, p. 7129
4. G.-S. Kim, I.-S. Ahn, S.-K. Oh, The design of optimized fuzzy neural networks and its application. Trans. Korean Inst. Electr. Eng. **58**(6), 1615–1623 (2009)
5. R. Martinez, A. Rodriguez, O. Castillo, L.T. Aguilar, Type-2 fuzzy logic controllers optimization using genetic algorithms and particle swarm optimization, in *Proceedings of the IEEE International Conference on Granular Computing, GrC 2010*, 2010, pp. 724–727
6. W.-H.R. Jeng, C.-Y. Yeh, S.-J. Lee, General type-2 fuzzy neural network with hybrid learning for function approximation, in *Proceedings of the IEEE Conference on Fuzzy Systems*, Jeju, Korea, 2009, pp. 1534–1539
7. R. Martinez, O. Castillo, L.T. Aguilar, A. Rodriguez, Optimization of type-2 fuzzy logic controllers using PSO applied to linear plants. Stud. Comput. Intell. **318**, 181–193 (2010)
8. M.A. Khanesar, M. Teshnehlab, E. Kayacan, O. Kaynak, A novel type-2 fuzzy membership function: Application to the prediction of noisy data, in *Proceedings of the IEEE International Conference on Computational Intelligence for Measurement Systems and Applications, CIMSA 2010*, 2010, pp. 128–133
9. Z. Bingül, O. Karahan, A fuzzy logic controller tuned with PSO for 2 DOF robot trajectory control. Expert Syst. Appl. **38**(1), 1017–1031 (2011)
10. S.-K. Oh, H.-J. Jang, W. Pedrycz, A comparative experimental study of type-1/type-2 fuzzy cascade controller based on genetic algorithms and particle swarm optimization. Expert Syst. Appl. **38**(9), 11217–11229 (2011)

Chapter 6
Ant Colony Optimization Algorithms for the Design of Type-2 Fuzzy Systems

There have also been several works reported in the literature optimizing type-2 fuzzy systems using different kinds of Ant Colony Optimization (ACO) algorithms. Most of these works have had relative success according to the different areas of application. In this chapter, we offer a representative and brief review of these types of works to illustrate the advantages of using the ACO optimization techniques for automating the design process or parameters of type-2 fuzzy systems.

In the work of Juang et al. [1], a Reinforcement Self-Organizing Interval Type-2 Fuzzy System with Ant Colony Optimization (RSOIT2FS-ACO) method was proposed. The antecedent part in each fuzzy rule of the RSOIT2FS-ACO uses interval type-2 fuzzy sets in order to improve system robustness to noise. The consequent part of each fuzzy rule was designed using the ACO technique. The ACO approach selects the consequent part from a set of candidate actions according to ant pheromone trails. The RSOIT2FS-ACO method was applied to a truck-backing control. The proposed RSOIT2FS-ACO was compared with other reinforcement fuzzy systems to verify its efficiency and effectiveness. A comparison with type-1 fuzzy systems verifies the robustness of using type-2 fuzzy systems to noise.

In the work of Martinez-Marroquin et al. [2], the application of a simple ACO as an optimization method for the membership functions' parameters of a fuzzy logic controller was proposed. The application of ACO enables finding the optimal intelligent controller for an autonomous wheeled mobile robot. In the ACO implementation, each interval type-2 fuzzy controller was represented as a trajectory on a graph. Simulation results show that ACO outperforms a GA in the optimization of interval type-2 fuzzy logic controllers for a particular autonomous wheeled mobile robot.

In the work of Juang and Hsu [3], a reinforcement ant optimized fuzzy controller (FC) design method, called RAOFC, was proposed. The method was applied it to wheeled mobile robot wall-following control under reinforcement-learning

O. Castillo and P. Melin, *Recent Advances in Interval Type-2 Fuzzy Systems*,
SpringerBriefs in Computational Intelligence, DOI: 10.1007/978-3-642-28956-9_6,

environments. The inputs to the designed FC are range-finding sonar sensors, and the controller output is a robot steering angle. The antecedent part in each fuzzy rule uses interval type-2 fuzzy sets in order to increase FC robustness. No a priori assignment of fuzzy rules was necessary in RAOFC. An online aligned interval type-2 fuzzy clustering (AIT2FC) method was proposed to generate rules automatically. The AIT2FC not only flexibly partitions the input space but also reduces the number of fuzzy sets in each input dimension, which improves controller interpretability. The consequent part of each fuzzy rule is designed using Q-value aided ant colony optimization (QACO). The QACO approach selects the consequent part from a set of candidate actions according to ant pheromone trails and Q-values, both of whose values are updated using reinforcement signals. Simulations and experiments on mobile-robot wall-following control show the effectiveness and efficiency of the proposed RAOFC.

In the work of Castillo et al. [4], the application of ACO and PSO for the optimization of an interval type-2 fuzzy logic controller for an autonomous wheeled mobile robot was presented. The obtained simulation results were statistically compared with the obtained previous work results achieved with GAs in order to determine the best optimization technique for this particular robotics problem. Both PSO and ACO were able to outperform GAs for this particular application. However, in comparing ACO and PSO, the best results were achieved with ACO. In this case, the authors claim that ACO is best suited for this particular robotic problem.

In the work of Juang and Hsu [5], a new reinforcement-learning method using Online Rule Generation and Q-value-aided Ant Colony Optimization (ORGQACO) for fuzzy controller design was proposed. The fuzzy controller is based on an interval type-2 fuzzy system (IT2FS). The antecedent part in the designed IT2FS uses interval type-2 fuzzy sets to improve controller robustness to noise. The ORGQACO concurrently designs both the structure and parameters of an IT2FS. An online interval type-2 rule generation method for the evolution of system structure and flexible partitioning of the input space was proposed. Consequent part parameters in an IT2FS are designed using Q-values and the reinforcement local–global ant colony optimization algorithm. This algorithm selects the consequent part from a set of candidate actions according to ant pheromone trails and Q-values, both of which are updated using reinforcement signals. The ORGQACO design method was applied to the following three control problems: (1) truck-backing control; (2) magnetic-levitation control; and (3) chaotic-system control. The ORGQACO was compared with other reinforcement-learning methods to verify its efficiency and effectiveness. Comparisons with type-1 fuzzy systems verify the noise robustness property of using an IT2FS.

In Table 6.1 a summary of the previously presented contributions, where ACO has been applied to optimize type-2 fuzzy systems, is presented. Table 6.1 shows that at the moment all the works have been done in the area of type-2 fuzzy logic controller design using different ACO methods. The comparison shown in Table 6.1 is based on the following criteria: author names, year of publication, reference number, domain of the problem, if a comparison with type-1 fuzzy logic

Table 6.1 ACO optimization of type-2 fuzzy systems

Author(s) (pub. Year)	Ref. no.	Domain of the problem	Comparison with type-1	Comparison with other optimization	Why type-2 is required for the problem?
Juang et al. 2009	[1]	Control	Yes	No	Uncertainty in control
Martinez-Marroquin et al. 2009	[2]	Control	Yes	Yes	Uncertainty in mobile robots
Juang and Hsu 2009	[3]	Control	No	No	Uncertainty in navigation
Castillo et al. 2011	[4]	Control	Yes	Yes	Modeling uncertainty in control
Juang and Hsu 2009	[5]	Control	Yes	No	Test different controllers

is provided, if a comparison with other optimization methods is presented, and why type-2 fuzzy logic was used by the authors. It can also be noted that at the moment, the number of papers mentioning the use of ACO is lower than the ones using PSO or GAs.

In conclusion, the use of ant colony algorithms for optimizing type-2 fuzzy systems is not so widespread yet, however we expect that in the future not only it will be used more in control it will also be used in pattern recognition, classification and time series prediction. The reason that we believe this to be true is that ACO has achieved very good results in the works reported in the literature, so this will hopefully encourage other researchers to work in this area.

References

1. C.-F. Juang, C.-H. Hsu, C.-F. Chuang, Reinforcement self-organizing interval type-2 fuzzy system with ant colony optimization, in *Proceedings of IEEE International Conference on Systems, Man and Cybernetics*, San Antonio, 2009, pp. 771–776
2. R. Martinez-Marroquin, O. Castillo, J. Soria, Parameter tuning of membership functions of a type-1 and type-2 fuzzy logic controller for an autonomous wheeled mobile robot using ant colony optimization, in *Proceedings of IEEE International Conference on Systems, Man and Cybernetics*, San Antonio, 2009, pp. 4770–4775
3. C.-F. Juang, C.-H. Hsu, Reinforcement ant optimized fuzzy controller for mobile-robot wall-following control. IEEE Trans. Ind. Electron. **56**(10), 3931–3940 (2009)
4. O. Castillo, R. Martinez-Marroquin, P. Melin, F. Valdez, J. Soria, Comparative study of bio-inspired algorithms applied to the optimization of type-1 and type-2 fuzzy controllers for an autonomous mobile robot. Inf. Sci. **192**(1), 19–38 (2012)
5. C.-F. Juang, C.-H. Hsu, Reinforcement interval type-2 fuzzy controller design by online rule generation and Q-value-aided ant colony optimization. IEEE Trans. Syst. Man Cybern. B Cybern. **39**(6), 1528–1542 (2009)

Chapter 7
Other Methods for Optimization of Type-2 Fuzzy Systems

In this chapter we describe some other works reported in the literature optimizing type-2 fuzzy systems using different kinds of optimization algorithms (other than GAs, PSO or ACO, which were covered in previous chapters). Most of these works have had relative success according to the different areas of application. In this chapter, we offer a representative and brief review of these types of works to illustrate the advantages of using the corresponding optimization techniques for automating the design process or parameters of type-2 fuzzy systems.

In the work by Aliev et al. [1], type-2 fuzzy neural networks with fuzzy clustering and differential evolution are presented. Type-2 fuzzy logic systems developed with the aid of evolutionary optimization forms a useful modeling tool subsequently resulting in a collection of efficient "If–Then" rules. Type-2 fuzzy neural networks take advantage of capabilities of fuzzy clustering by generating type-2 fuzzy rule base, resulting in a small number of rules and then optimizing membership functions of type-2 fuzzy sets present in the antecedent and consequent parts of the rules. The clustering itself is realized with the aid of differential evolution. Several examples, including a benchmark problem of identification of nonlinear system, are considered. The reported comparative analysis of experimental results was used to quantify the performance of the developed networks.

In the work by Hidalgo et al. [2, 3], a method for the optimization of type-2 fuzzy systems based on the level of uncertainty, considering three different cases to reduce the complexity problem of searching the solution space, was presented. The proposed method produces the best fuzzy inference systems for particular applications with the help of a genetic algorithm. The application of a genetic algorithm to find the optimal type-2 fuzzy system is performed by dividing the search space in three subspaces. The division is performed by considering three different cases in the design of type-2 fuzzy systems: (1) equal membership functions, (2) equal membership functions in each variable, and (3) different membership functions for all the linguistic values. Comparative results and analysis for the benchmark problems were produced with good success.

O. Castillo and P. Melin, *Recent Advances in Interval Type-2 Fuzzy Systems*,
SpringerBriefs in Computational Intelligence, DOI: 10.1007/978-3-642-28956-9_7,

In the work by Mohammadi et al. [4], an evolutionary tuning technique for type-2 fuzzy logic controller was presented. Uncertainty is an inherent part in control systems used in real world applications. Various instruments used in such systems produce uncertainty in their measurements and thus influence the integrity of the data collection. Type-1 fuzzy sets used in conventional fuzzy systems cannot fully handle the uncertainties present but type-2 fuzzy sets that are used in type-2 fuzzy systems can handle such uncertainties in a better way because they provide more parameters and more design degrees of freedom. There are membership functions which can be parameterized by a few variables and when optimized, the membership optimization problem can be reduced to a parameter optimization problem. This work deals with the parameter optimization of the type-2 fuzzy membership functions using a new proposed reinforcement learning algorithm in a nonlinear system. The results of the proposed method referred to as Extended Discrete Action Reinforcement Learning Automata algorithm were compared to the results obtained by the Discrete Action Reinforcement Learning Automata and Continuous Action Reinforcement Learning Automata algorithms. The Performance of the proposed method on initial error reduction and error convergence issues were also investigated by computer simulations.

In the work by Hidalgo et al. [5, 6], an evolutionary method for the optimization of type-2 fuzzy systems based on the level of uncertainty was proposed. The proposed evolutionary method produces the best fuzzy inference systems (based on the membership functions) for particular applications. The optimization of membership functions of the type-2 fuzzy systems is based on the level of uncertainty considering three different cases to reduce the complexity problem of searching the solution space. The method for optimizing type-2 fuzzy systems was applied to find the optimal integration method for Modular Neural Networks (MNN) in pattern recognition.

In the work of Castillo et al. [7], a method for designing optimal interval type-2 fuzzy logic controllers using evolutionary algorithms was presented. Interval type-2 fuzzy controllers can outperform conventional type-1 fuzzy controllers when the problem has a high degree of uncertainty. However, designing interval type-2 fuzzy controllers is more difficult because there are more parameters involved. In this work, interval type-2 fuzzy systems were approximated with the average of two type-1 fuzzy systems, which has been shown to give good results in control if the type-1 fuzzy systems can be obtained appropriately. An evolutionary algorithm is applied to find the optimal interval type-2 fuzzy system as mentioned above. The human evolutionary model is applied for optimizing the interval type-2 fuzzy controller for a particular non-linear plant and results were compared against an optimal type-1 fuzzy controller. A comparative study of simulation results of the type-2 and type-1 fuzzy controllers, under different noise levels, was also presented. Simulation results show that interval type-2 fuzzy controllers obtained with the evolutionary algorithm outperform type-1 fuzzy controllers.

In the work by Garcia [8], a case study of the US Dollar-Colombian Peso Exchange Rate by using a First order Interval Type-2 TSK Fuzzy Logic System was presented. This case study is especially interesting because it presents a volatile behavior, which is a complex problem for classical analysis. The results were verified by statistical tests, such as Bayesian, Akaike, Hannan-Quin criteria, Goldfeld-Quant, Ljung-Box, ARCH, Runs and Turning Points which provide appropriate criterions to test the solution. Some methodological aspects about the hybrid implementation combining evolutionary optimization and first order Interval Type-2 TSK FLS were presented.

In the work by Muñoz et al. [9], the development of fuzzy response integrators for a MNN and its Optimization with a Hierarchical Genetic Algorithm (HGA) was presented. The optimization of the integrators consists of optimizing their membership functions, fuzzy rules, type of model (Mamdani or Sugeno), and type of fuzzy logic (type-1 or type-2). The MNN architecture consists of three modules; face, fingerprint and voice. Each of the modules is divided again into three sub modules. The same information is used as input to train the sub modules. In this work the use of HGAs as optimization techniques for the fuzzy integrators is shown to be a good option to solve the MNN integration problems.

In the work by Menolascina et al. [10], a method for induction of fuzzy rules based on artificial immune systems was proposed. Fuzzy Rule Induction (FRI) is one of the main areas of research in the field of computational intelligence. Recently FRI has been successfully employed in the field of data mining in bioinformatics. Thanks to its flexibility and potentialities FRI allowed researchers to extract rules that can be easily modeled in natural language and submitted to experts in the field that can validate their accuracy or consistency. The process of FRI can result to be highly complex from a computational complexity point of view and, for this reason, several alternative approaches to accomplish this process have been proposed ranging from iterative and simultaneous algorithms to GAs and ACO based approaches. This work focuses on a specific application of type-1 (T1) and type-2 (T2) fuzzy systems to data mining in bioinformatics in which FRI is carried out using a novel and promising computational paradigm, namely Artificial Immune Systems (AIS).

In the work by Poornaselvan et al. [11], the main objective was to focus on an agent based approach to flight control in ground/runway. The idea was to provide an autonomous control on flight once the airplane comes to runway. In all airports there is a particular structure for the runway, like main runways, sub runways, different tracks. An interval type 2 fuzzy controller can be applied to the autonomous vehicle in order to handle uncertainty in a better way. Ant colony optimization technique can be used for an optimized path planning in traffic environment with more number of flights. A hybrid ant colony optimization was used to handle real time dynamic environment and path planning. Both Agent based and type-2 fuzzy logic together with the ACO technique were used to achieve another level of intelligence.

In the work by Castillo et al. [12], an evolutionary computing based approach for the optimization of type-2 fuzzy systems was presented. In particular, the application of HGAs for fuzzy system optimization in intelligent control was proposed. The problem of optimizing the number of rules and membership functions using an evolutionary approach was considered. The HGA enables the optimization of the fuzzy system design for a particular application. The proposed approach was illustrated with the case of intelligent control in a medical application. Simulation results for this application show that an optimal set of rules and membership functions for the fuzzy system can be obtained with this approach.

In the work by Astudillo et al. [13], an optimization method based on the chemical reaction paradigm was proposed. The new optimization method was inspired on a nature based paradigm: the reaction methods existing on chemistry, and the way the elements combine with each other to form compounds, in other words, quantum chemistry. The proposed optimization method was tested with benchmark mathematical functions and also with the design of type-2 fuzzy controllers with excellent results.

In the work by Castillo et al. [14], the use of HGAs for type-2 fuzzy system optimization in anesthesia control was proposed. In particular, the problem of optimizing the number of rules and membership functions using an evolutionary approach was proposed. The HGA enables the optimization of the type-2 fuzzy system design for the particular application of anesthesia control. Simulation results for this application show that an optimal set of rules and membership functions for the type-2 fuzzy controller can be obtained in an efficient manner.

The work by Astudillo et al. [15], focuses on the control of wheeled mobile robot under bounded torque disturbances. A hybrid tracking controller for the mobile robot was developed by considering its kinematic model and Euler–Lagrange dynamics. The procedure consists in minimizing the stabilization error of the kinematic model through a genetic algorithm approach while attenuation to perturbed torques is made through type-2 fuzzy logic control via backstepping methodology. Type-2 fuzzy logic was proposed to synthesize the controller for the overall system, which is claimed to be a robust tool for related applications. The theoretical results were illustrated through computer simulations of the closed-loop system.

In the work by Sepulveda et al. [16], a method for optimizing type-2 fuzzy logic controllers based on the human evolutionary model is proposed. The automatic design of optimal type-2 fuzzy controllers is performed using an efficient evolutionary algorithm based on human characteristics. The human evolutionary model produces interval type-2 fuzzy controllers, for benchmark non-linear plants, that are shown to perform better than the ones obtained with other well known optimization algorithms.

In Table 7.1 a summary of the contributions where other optimization methods (different than GAs, PSO and ACO) have been applied to design type-2 fuzzy systems is presented. The comparison shown in Table 7.1 is based on the domain of the problem, if a comparison with type-1 fuzzy logic is provided, if a comparison with other optimization methods is presented, and why type-2 fuzzy logic was used by the authors. From Table 7.1 it can be noted that most of the applications have been developed in the intelligent control area with only a few applications in pattern recognition, and time series prediction.

In this chapter a review of different optimization methods (other than GAs, PSO and ACO) that have been applied to the optimization of type-2 fuzzy systems is

Table 7.1 Other methods of optimization for type-2 fuzzy systems

Author(s) (pub. Year)	Ref. no.	Domain of the problem	Comparison with type-1	Comparison with other optimization	Why type-2 is required for the problem?
Aliev et al. 2011	[1]	System identification	Yes	Yes	Uncertainty in identification
Hidalgo et al. 2009	[2, 3]	Prediction	Yes	No	Uncertainty in prediction
Mohammadi et al. 2010	[4]	Control	Yes	No	Uncertainty in control
Hidalgo et al. 2010	[5, 6]	Prediction	Yes	No	Uncertainty in prediction
Castillo et al. 2011	[7]	Control	Yes	No	Comparison of controllers
Garcia 2009	[8]	Prediction	No	No	Noise in prediction
Munoz et al. 2009	[9]	Recognition	Yes	No	Uncertainty in recognition
Menolascina et al. 2009	[10]	Data mining	Yes	Yes	Uncertainty in rule induction
Poornaselvan et al. 2008	[11]	Control	No	No	Uncertainty in control
Castillo et al. 2007	[12]	Control	Yes	No	Uncertainty in control
Astudillo et al. 2010	[13]	Control	Yes	Yes	Uncertainty in control
Castillo et al. 2008	[14]	Control	Yes	No	Uncertainty in control
Astudillo et al. 2007	[15]	Control	Yes	No	Uncertainty in mobile robots
Sepulveda et al. 2011	[16]	Control	Yes	No	Uncertainty in control

presented. Other optimization methods, like simulated annealing, bee optimization, harmony search, tabu search, etc., are not mentioned in this review because to the moment there are no works reported in the literature on optimizing type-2 fuzzy systems with these optimization methods. However, we expect that in the near future these methods and most likely others that may arise, would be applied to the problem of type-2 fuzzy system optimization.

References

1. R.A. Aliev, W. Pedrycz, B.G. Guirimov, R.R. Aliev, U. Ilhan, M. Babagil, S. Mammadli, Type-2 fuzzy neural networks with fuzzy clustering and differential evolution optimization. Inf. Sci. **181**(9), 1591–1608 (2011)
2. D. Hidalgo, P. Melin, O. Castillo, Type-2 fuzzy inference system optimization based on the uncertainty of membership functions applied to benchmark problems. Lecture Notes in Computer Science, vol. 6438 (2010), pp. 454–464
3. D. Hidalgo, P. Melin, O. Castillo, Optimal design of type-2 fuzzy membership functions using genetic algorithms in a partitioned search space, in *Proceedings of the IEEE International Conference on Granular Computing, GrC 2010*, San Jose, Aug 2010, pp. 212–216
4. S.M.A. Mohammadi, A.A. Gharaveisi, M. Mashinchi, An evolutionary tuning technique for type-2 fuzzy logic controller in a non-linear system under uncertainty, in *Proceedings of the 18th Iranian Conference on Electrical Engineering, ICEE 2010*, pp. 610–616
5. D. Hidalgo, P. Melin, G. Licea, O. Castillo, Optimization of type-2 fuzzy integration in modular neural networks using an evolutionary method with applications in multimodal biometry. Lecture Notes in Computer Science, vol. 5845 (2009), pp. 454–465
6. D. Hidalgo, P. Melin, O. Mendoza, Evolutionary optimization of type-2 fuzzy systems based on the level of uncertainty, in *Proceedings of the IEEE World Congress on Computational Intelligence, WCCI 2010*, Barcelona, July 2010
7. O. Castillo, P. Melin, A. Alanis, O. Montiel, R. Sepulveda, Optimization of interval type-2 fuzzy logic controllers using evolutionary algorithms. J. Soft Comput. **15**(6), 1145–1160 (2011)
8. J.C.F. Garcia, An evolutive interval type-2 TSK fuzzy logic system for volatile time series identification, in *Proceedings of the IEEE International Conference on Systems, Man and Cybernetics*, 2009, pp. 666–671
9. R. Muñoz, O. Castillo, P. Melin, Optimization of fuzzy response integrators in modular neural networks with hierarchical genetic algorithms: the case of face, fingerprint and voice recognition. Stud. Comput. Intell. **257**, 111–129 (2009)
10. F. Menolascina, V. Bevilacqua, M. Zarrilli, G. Mastronardi, Induction of fuzzy rules by means of artificial immune systems in bioinformatics. Stud. Fuzziness Soft Comput. **242**, 1–17 (2009)
11. K.J. Poornaselvan, T. Gireesh Kumar, V.P. Vijayan, Agent based ground flight control using type-2 fuzzy logic and hybrid ant colony optimization to a dynamic environment, in *Proceedings of the 1st International Conference on Emerging Trends in Engineering and Technology, ICETET 2008*, 2008, pp. 343–348
12. O. Castillo, A.I. Martinez, A.C. Martinez, Evolutionary computing for topology optimization of type-2 fuzzy systems. Adv. Soft Comput. **41**, 63–75 (2007)
13. L. Astudillo, P. Melin, O. Castillo, A new optimization method based on a paradigm inspired by nature. Stud. Comput. Intell. **312**, 277–283 (2010)
14. O. Castillo, G. Huesca, F. Valdez, Evolutionary computing for topology optimization of type-2 fuzzy controllers. Stud. Fuzziness Soft Comput. **208**, 163–178 (2008)

15. L. Astudillo, O. Castillo, L.T. Aguilar, R. Martinez, Hybrid control for an autonomous wheeled mobile robot under perturbed torques. Lecture Notes in Computer Science, vol. 4529 (2007), pp. 594–603
16. R. Sepulveda, O. Montiel, O. Castillo, P. Melin, Optimizing the MFs in type-2 fuzzy logic controllers, using the human evolutionary model. Int. Rev. Autom. Control **3**(1), 1–10 (2011)

Chapter 8
Simulation Results Illustrating the Optimization of Type-2 Fuzzy Controllers

In this chapter we describe as an illustration the optimization of the membership functions' (MF) parameters of a type-2 fuzzy logic controller (FLC) in order to find the optimal intelligent controller for an autonomous wheeled mobile robot. The complete details of the robot, the fuzzy controller and simulation results can be found in [1].

The model considered is that of a unicycle mobile robot (see Fig. 8.1) that has two driving wheels fixed to the axis and one passive orientable wheel that is placed in front of the axis and normal to it [2].

The two fixed wheels are controlled independently by the motors, and the passive wheel prevents the robot from overturning when moving on a plane.

It is assumed that the motion of the passive wheel can be ignored from the dynamics of the mobile robot, which is represented by the following set of equations [3]:

$$\dot{q} = \begin{vmatrix} \cos\theta & 0 \\ \sin\theta & 0 \\ 0 & 1 \end{vmatrix} \begin{vmatrix} v \\ w \end{vmatrix} \tag{8.1}$$

$$M(q)\dot{v} + V(q,\dot{q})v + G(q) = \tau \tag{8.2}$$

Where $q = [x, y, \theta]^T$ is the vector of generalized coordinates which describes the robot position, *(x,y)* are the Cartesian coordinates, which denote the mobile center of mass and θ is the angle between the heading direction and the *x*-axis (which is taken counterclockwise form); $v = [v, w]^T$ is the vector of velocities, v and w are the linear and angular velocities respectively; $\tau \in R^r$ is the input vector, $M(q) \in R^{n \times n}$ is a symmetric and positive-definite inertia matrix, $V(q,\dot{q}) \in R^{n \times n}$ is the centripetal and Coriolis matrix, $G(q) \in R^n$ is the gravitational vector. Equation (8.1) represents the kinematics or steering system of a mobile robot.

O. Castillo and P. Melin, *Recent Advances in Interval Type-2 Fuzzy Systems*,
SpringerBriefs in Computational Intelligence, DOI: 10.1007/978-3-642-28956-9_8,

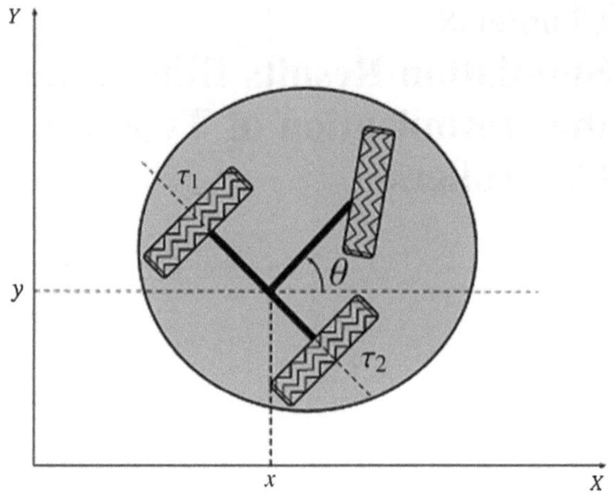

Fig. 8.1 Wheeled mobile robot

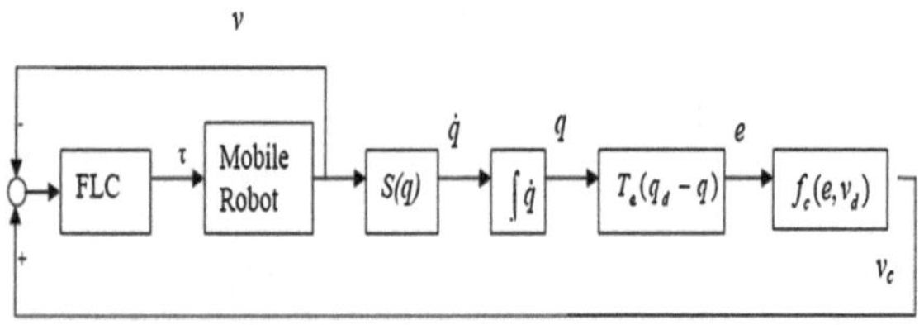

Fig. 8.2 Tracking control structure

Notice that the no-slip condition imposed a non holonomic constraint described by (8.3), that it means that the mobile robot can only move in the direction normal to the axis of the driving wheels.

$$\dot{y}\cos\theta - \dot{x}\sin\theta = 0 \tag{8.3}$$

8.1 Tracking Controller of Mobile Robot

The control objective will be established as follows: Given a desired trajectory $q_d(t)$ and the orientation of the mobile robot we must design a controller that applies an adequate torque τ such that the measured positions $q(t)$ achieve the desired reference $q_d(t)$ represented as (8.4):

$$\lim_{t\to\infty} \|q_d(t) - q(t)\| = 0 \tag{8.4}$$

To reach the control objective, we are based on the procedure of [2], we are deriving a $\tau(t)$ of a specific $v_c(t)$ that controls the steering system (8.1) using a FLC. A general structure of tracking control system is presented in Fig. 8.2.

8.2 Control of the Kinematic Model

We are based on the procedure proposed by Astudillo et al. [2] and to solve the tracking problem for the kinematic model $v_c(t)$. Suppose that the desired trajectory q_d satisfies (8.5):

$$\dot{q}_d = \begin{vmatrix} \cos\theta_d & 0 \\ \sin\theta_d & 0 \\ 0 & 1 \end{vmatrix} \begin{vmatrix} v_d \\ w_d \end{vmatrix} \tag{8.5}$$

Using the robot local frame (the moving coordinate system x–y in Fig. 8.1), the error coordinates can be defined as (8.6):

$$e = T_e(q_d - q), \begin{vmatrix} e_x \\ e_y \\ e_\theta \end{vmatrix} = \begin{vmatrix} \cos\theta & \sin\theta & 0 \\ -\sin\theta & \cos\theta & 0 \\ 0 & 0 & 1 \end{vmatrix} \begin{vmatrix} x_d - x \\ y_d - y \\ \theta_d - \theta \end{vmatrix} \tag{8.6}$$

And the auxiliary velocity control input that achieves tracking for (8.1) is given by (8.7):

$$v_c = f_c(e, v_d), \begin{vmatrix} v_c \\ w_c \end{vmatrix} = \begin{vmatrix} v_d + \cos e_\theta + k_1 e_x \\ w_d + v_d k_2 e_y + v_d k_3 \sin e_\theta \end{vmatrix} \tag{8.7}$$

Where k_1, k_2 and k_3 are positive gain constants.

The first approach in the optimization for the fuzzy controller of the mobile robot is to apply our proposed method to obtain the values of k_i (i = 1, 2, 3) for optimal behavior of the controller.

Once we have found these gain constraints, the second step is to find the values of the MF of the fuzzy controller.

8.3 The Fuzzy Logic Tracking Controller

The purpose of the FLC is to find a control input τ such that the current velocity vector v is able to reach the velocity vector v_c this is denoted as (8.8):

$$\lim_{t\to\infty} \|v_c - v\| = 0 \tag{8.8}$$

The inputs variables of the FLC correspond to the velocity errors obtained of (8.6) (denoted as *ev* and *ew*: linear and angular velocity errors respectively), and 2 outputs variables, the driving and rotational input torques τ (denoted by F and N respectively).

The initial MF are defined by 1 triangular and 2 trapezoidal functions for each variable involved. In future work the shape of the MF will be selected by the algorithm as part of the optimization.

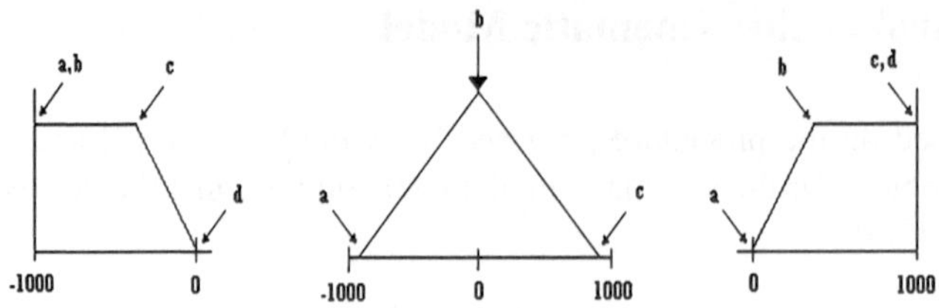

Fig. 8.3 Membership functions of the input/output variables

Table 8.1 Upper and lower limits of the membership functions

Membership function	Point	Lower limit	Upper limit
Trapezoidal	a	−1000	−1000
	b	−1000	−1000
	c	−800	−300
	d	−300	250
Triangular	a	−800	−200
	b	−50	50
	c	200	800
Trapezoidal	a	−250	300
	b	300	800
	c	1000	1000
	d	1000	1000

Table 8.2 Fuzzy rule set

e_v/e_w	N	C	P
N	N/N	N/Z	N/P
Z	Z/N	Z/Z	Z/P
P	P/N	P/Z	P/P

Figure 8.3 depicts the MFs in which N, C, P represent the fuzzy sets (Negative, Zero and Positive respectively) associated to each input and output variable.

Table 8.1 shows the upper and lower limits of the used MF.

The rule set of the FLC contain nine rules, which govern the input–output relationship of the FLC and this adopts the Mamdani-style inference engine. We use the center of gravity method to realize defuzzification procedure. In Table 8.2, we present the rule set whose format is established as follows:

$$\text{Rule i: If ev is G1 and ew is G2 then F is G3 and N is G4} \tag{8.9}$$

Where G1..G4 are the fuzzy set associated to each variable and i = 1… 9.

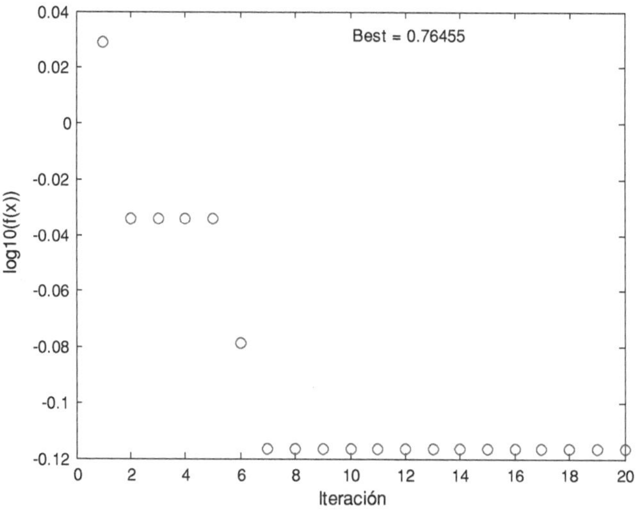

Fig. 8.4 Optimization behavior for the S-ACO on Type-2 FLC optimization

8.4 Control of an Autonomous Mobile Robot Using Type-2 Fuzzy Logic

The tracking controller obtained by means of fuzzy logic was considered as a base to design a type-2 FLC.

The membership function types and parameters of the primary MF are the same that resulted in the type-1 fuzzy controller.

The parameters that the chemical reaction paradigm will attempt to find are those for the secondary membership function.

Once these parameters are found, the objective is to test the performance of the FLCs (type-1 and type-2) by applying a perturbance to the tracking controller system that is given by:

$$\text{Perturbation} = \varepsilon \sin \omega t \tag{8.10}$$

Where t = time, in an interval of 1–10 s, $\varepsilon = 0.001$ and $\omega = 1$.

Figure 8.4 shows the optimization behavior of the ACO method. Figure 8.5 shows the MF of the FLC obtained by the simple ACO algorithm. Figure 8.6 shows both the desired trajectory and obtained trajectory for the robot.

In this application a trajectory tracking controller was designed based on the dynamics and kinematics of the autonomous mobile robot through the application of ACO for the optimization of MF for the type-1 and type-2 fuzzy controllers with good results obtained after simulations. The optimal type-2 fuzzy controller was able to outperform the best type-1 fuzzy controller. Also, simulation results in this application provided sufficient statistical information to say that ACO outperforms

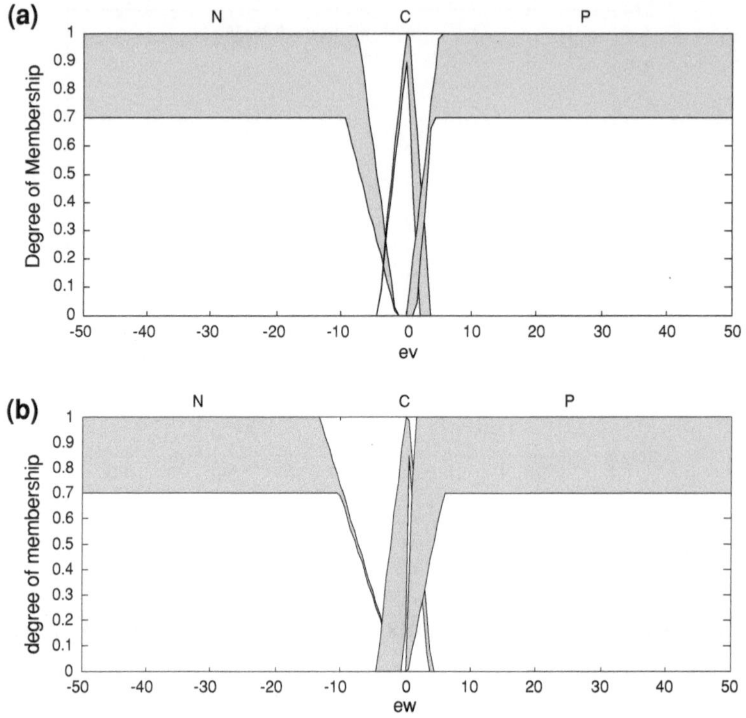

Fig. 8.5 Membership functions: **a** linear velocity error, **b** angular velocity error optimized by the ACO algorithm

(on average) PSO and the GA, but PSO outperforms the GA. In any case, the three optimization methods are able to optimize the fuzzy controllers (to a certain level) and the difference is in the achieved tracking errors, which are lower for the ACO optimization method [1].

8.5 Results of the CRA Applied to the Fuzzy Control of an Autonomous Mobile Robot

8.5.1 *Finding* k_1, k_2, k_3

The first approach in the optimization for the fuzzy controller of the mobile robot was to apply our proposed method to obtain the values of k_i (i = 1, 2, 3) for optimal behavior of the controller.

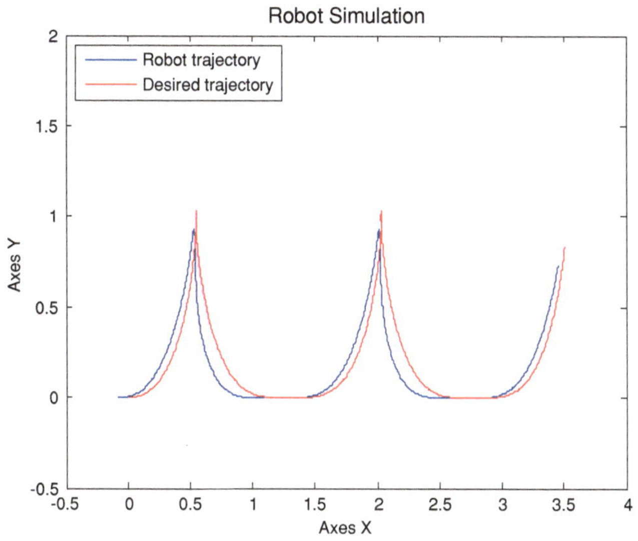

Fig. 8.6 Obtained trajectories with type-2 FLC optimization

Table 8.3 Parameters of the chemical reaction optimization

A	B	C	D	E
1	2	10	2	0.3
2	5	10	3	0.3
3	2	10	2	0.4
4	2	10	3	0.4
5	5	10	2	0.4
6	5	10	3	0.4
7	5	10	2	0.5
8	10	10	2	0.5

A, Identification number for each experiment; B, Initial elements—Initial pool of compounds randomly created; C, Trials—Number of iterations per experiment; D, Decomposition rate—Percentage of compounds to be decomposed; E, Decomposed elements—Number of elements resulted from applying the decomposition reaction

Several tests of the chemical optimization paradigm were made to test the performance of the tracking controller. The test parameters can be observed in Table 8.3. For statistical purposes, every experiment was executed 35 times.

The decomposed rate was considered to be an important parameter in this algorithm. Unlike previous bio-inspired optimization algorithms [4, 5] where the best individuals are selected to perform a genetic operation, this method applies the decomposition and composition reaction method to the worst compounds/elements of the pool, keeping the compounds/elements with better performance

Table 8.4 Experimental Results of the proposed method

A	B	C	D	k_1	k_2	k_3
1	0.0086	1.1568	3	519.86	46.52	8.85
2	4.79e−004	0.1291	5	205.81	31.05	31.05
3	0.0025	0.5809	7	36.06	328.61	88.68
4	0.0012	0.5589	8	2.76	206.18	0.37
5	0.0035	0.0480	2	185.19	29.92	5.11
6	8.13e−005	0.0299	3	270.35	53.68	15.02
7	0.0066	0.1440	4	29.25	15.94	0.027
8	0.0019	0.1625	8	51.93	3.69	0.001

A, Identification number for each experiment; B, Best error found; C, Mean of errors; D, Total trials of the experiment

through all the iterations, unless new elements/compounds with better performance are generated.

Following this criteria, for a pool containing 5 compounds, the quantity of compounds to compose and/or decompose is 2, if the decomposition rate is 0.4.

Table 8.4 shows the results after applying the chemical optimization paradigm.

Note that there was no need to increase the initial pool size of compounds,—which were randomly generated—, and this is because of the combination of the decomposition rate and the number of elements that every compound was decomposed.

That is, whenever some compound with poor fitness was found, it was a candidate to be decomposed in the next iteration. The decomposition was made by generating a random set of numbers between 0 and 1, and applying this factor to the original compound.

The value of the resultant elements must satisfy the following Eq. (8.11):

$$X = \sum_{i=1}^{n} x_i \tag{8.11}$$

Where X is the original compound, x is the resultant elements of the decomposition and i is the decomposition factor.

Figure 8.7 shows the behavior of the algorithm and the position errors in Simulink for the experiment No. 3, respectively, which was the best overall result so far, considering the average error and the positions error in x, y and *theta*.

In a previous work made by the authors [6], the gain constant values were found by means of genetic algorithms. Table 8.5 shows the best result of the experiments made and the obtained values for the gain constants using GAs.

Figure 8.8 shows the result in Simulink for the experiment with the best overall result, applying GAs as optimization method.

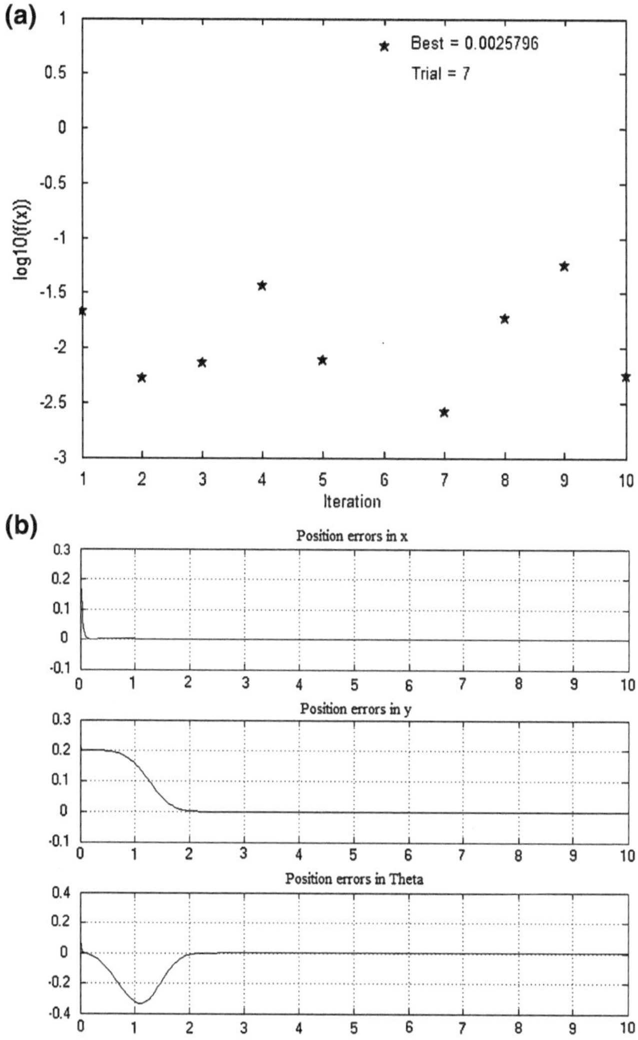

Fig. 8.7 **a** Convergence of the elements in experiment No. 3. **b** Final position errors achieved in experiment No. 3

Table 8.5 Best results using GAs

Error	k1	k2	k3
0.006734	43	493	19

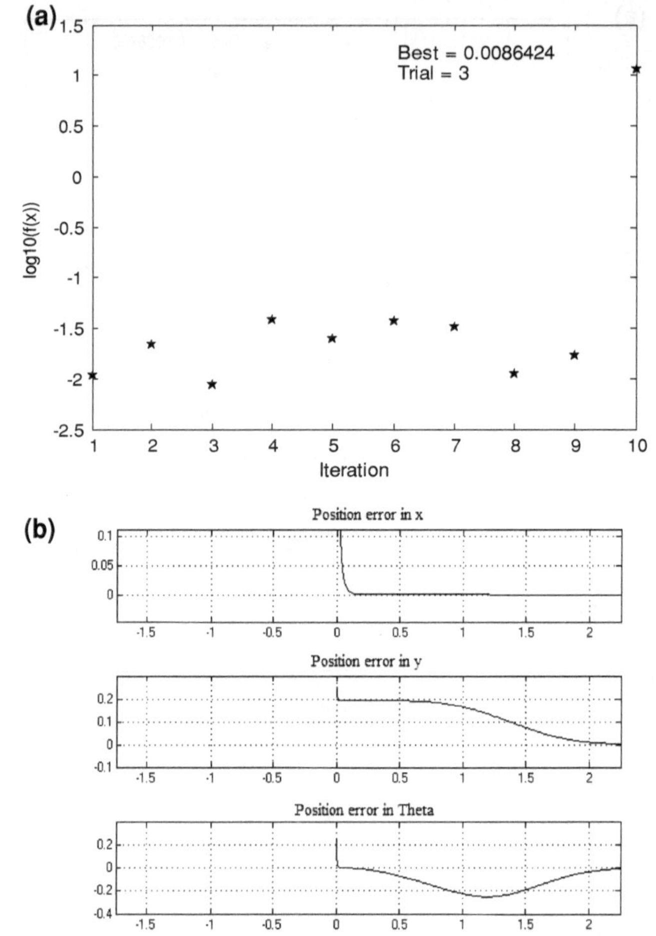

Fig. 8.8 **a** Convergence of the elements in experiment No. 1, using GAs. **b** Final position errors achieved in experiment No. 3, using GAs

8.5.2 Optimizing the Membership Function Parameters of the Fuzzy Controller

Once we have found optimal values for the gain constants, the next step is to find the optimal values for the input/output MF of the fuzzy controller. Our goal is that in the simulations, the linear and angular velocities reach zero.

The conditions for the simulations are shown in Eqs. (8.12–8.14). The expression of the desired trajectory is shown in Eq. (8.15), and Fig. 8.9 shows the control system designed in Simulink®.

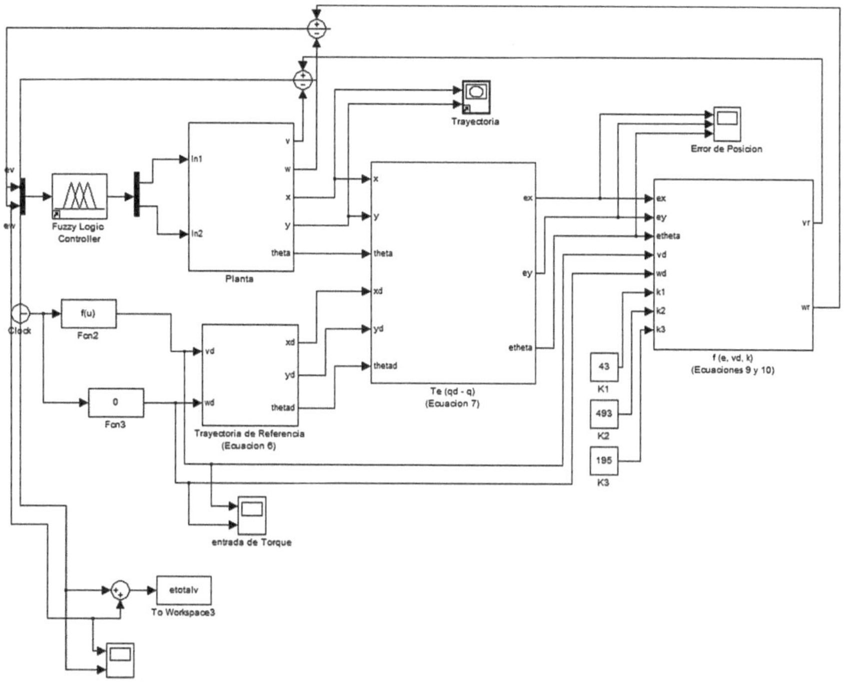

Fig. 8.9 Control system in Simulink®

Table 8.6 Parameters of the first set of simulations

Parameters	Value
Elements	10
Trials	15
Selection method	Stochastic universal sampling
K1	36
K2	328
K3	88
Error	0.077178

$$M(q) = \begin{bmatrix} 0.3749 & -0.0202 \\ -0.0202 & 0.3749 \end{bmatrix} \tag{8.12}$$

$$D = \begin{bmatrix} 10 & 0 \\ 0 & 10 \end{bmatrix} \tag{8.13}$$

$$C(q, \dot{q}) = \begin{bmatrix} 0 & 0.1350\dot{\theta} \\ 0.1350\dot{\theta} & 0 \end{bmatrix} \tag{8.14}$$

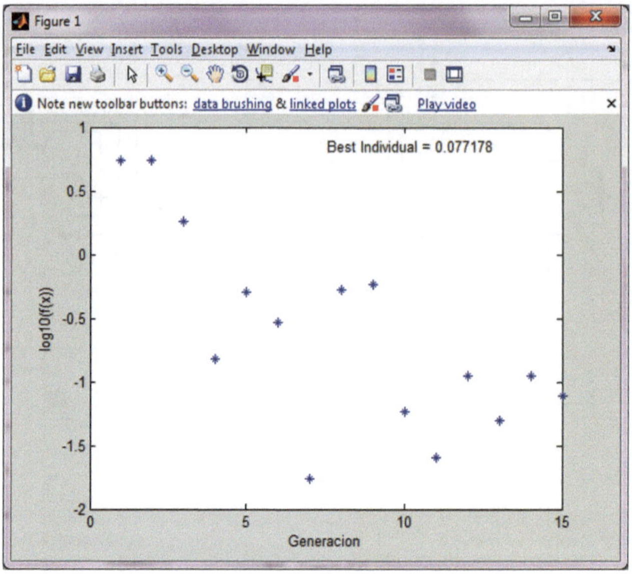

Fig. 8.10 Best simulation of experiment No. 1

$$v_d(t) = \left\{ \begin{matrix} v_d(t) = 0.25 - 0.25cos\left[\frac{2\pi t}{5}\right] \\ w_d(t) = 0 \end{matrix} \right\} \quad (8.15)$$

Table 8.6 shows the parameters used in the first set of simulations and Fig. 8.10 shows the behavior of the algorithm throughout the experiment.

Figure 8.11 shows the obtained input MF found by the proposed optimization algorithm.

Figure 8.12 shows the obtained output MF found by the proposed optimization algorithm.

Figure 8.13a shows the obtained trajectory when simulating the mobile control system including the obtained input and output MF; Fig. 8.13b shows the best trajectory reached by the mobile when optimizing the input and output MF using genetic algorithms.

8.6 Optimizing the Membership Function Parameters of the Type-2 Fuzzy Controller

The tracking controller obtained by means of fuzzy logic was considered as a base to design a type-2 FLC.

The membership function types and parameters of the primary MF are the same that resulted in the type-1 fuzzy controller.

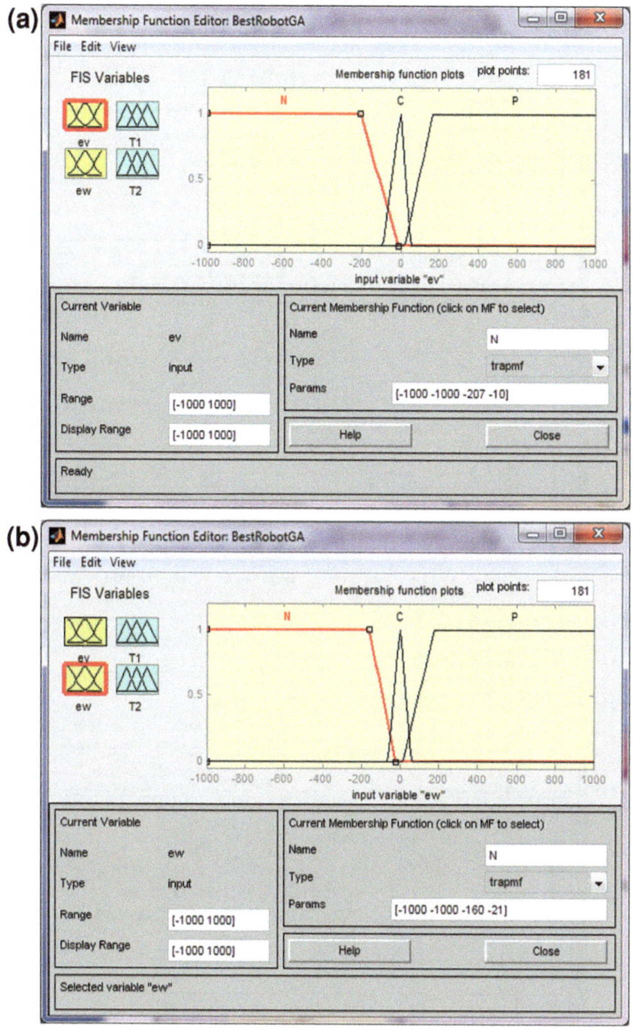

Fig. 8.11 Resulted input membership functions: **a** lineal and **b** angular velocities

The parameters that the chemical reaction paradigm will attempt to find are those for the secondary membership function. Table 8.7 shows the parameters used in the first set of simulations.

Figure 8.14 shows the behavior of the algorithm throughout the experiment.

Figures 8.15 and 8.16 show the obtained input and output MF found by the proposed optimization algorithm.

Figure 8.17 shows the obtained trajectory when simulating the mobile control system including the obtained input and output type-2 MF.

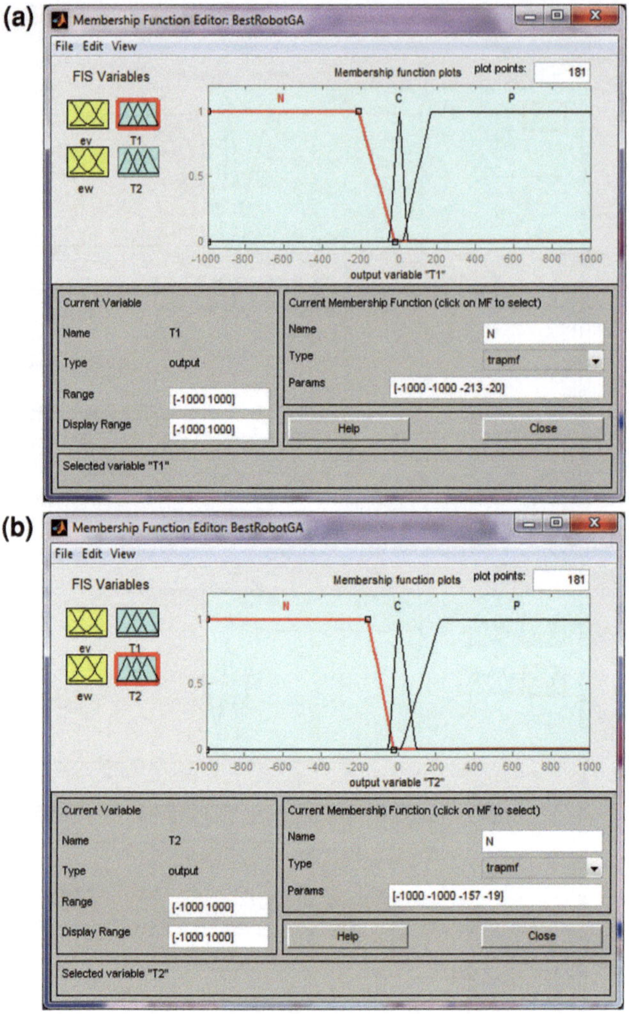

Fig. 8.12 Resulted output membership functions: **a** right and **b** left torque

In summary the proposed chemical optimization algorithm is able to obtain optimal parameter values for the fuzzy controller of the autonomous mobile robot. Comparison with alternative optimization methods also shows that the chemical optimization method is a good choice for this type of problems.

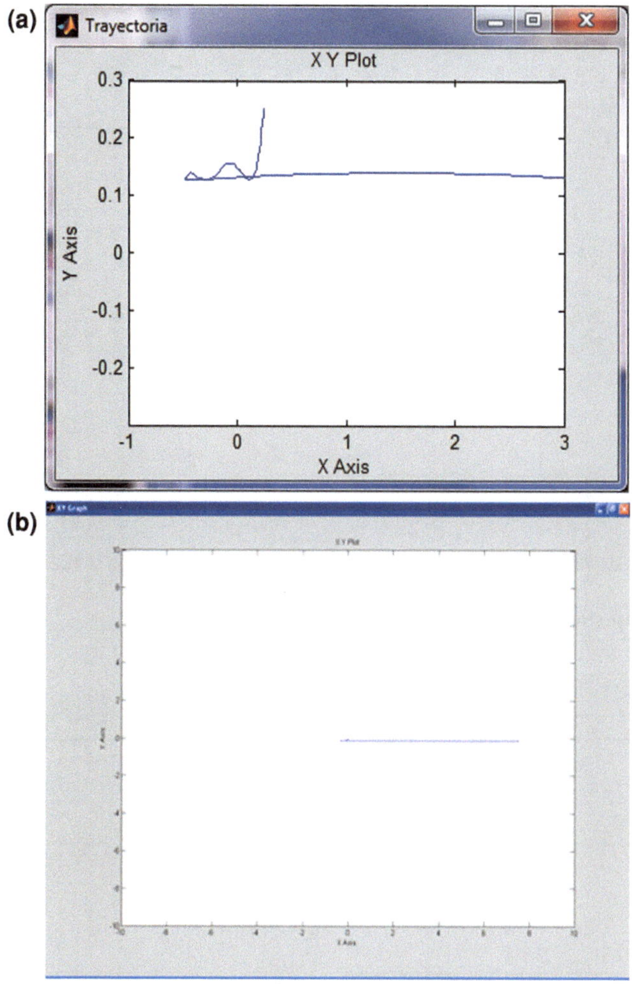

Fig. 8.13 **a** Obtained trajectory when applying the chemical reaction algorithm. **b** Obtained trajectory using genetic algorithms

Table 8.7 Parameters of the first set of simulations

Parameters	Value
Elements	10
Trials	10
Selection method	Stochastic universal sampling
K1	36
K2	328
K3	88
Error	2.7736

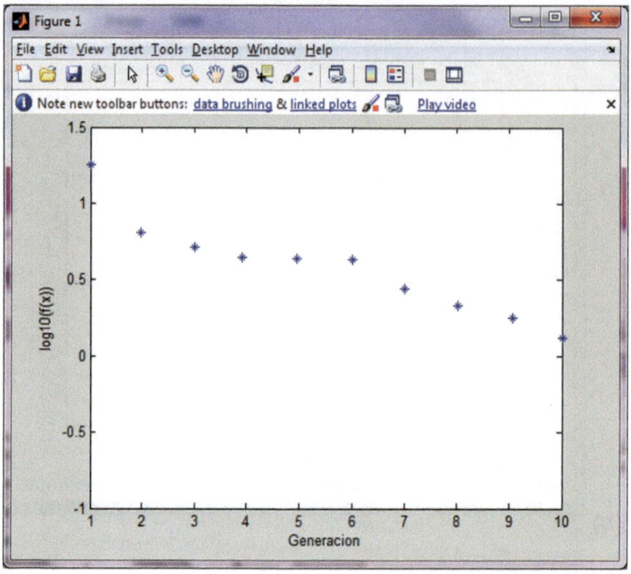

Fig. 8.14 Best simulation of experiment No. 1

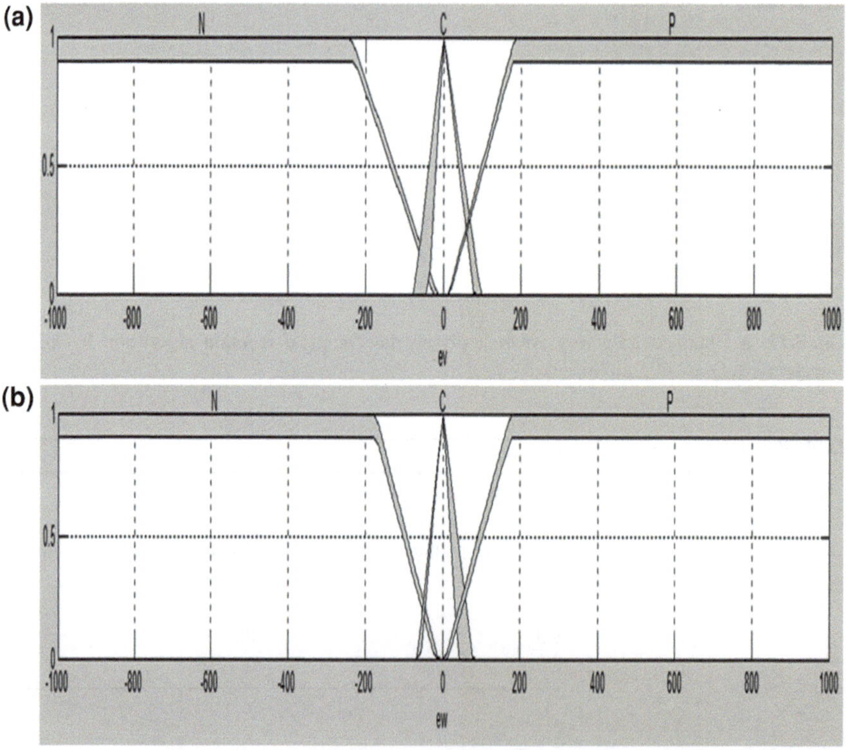

Fig. 8.15 Resulting input membership functions: **a** linear and **b** angular velocities

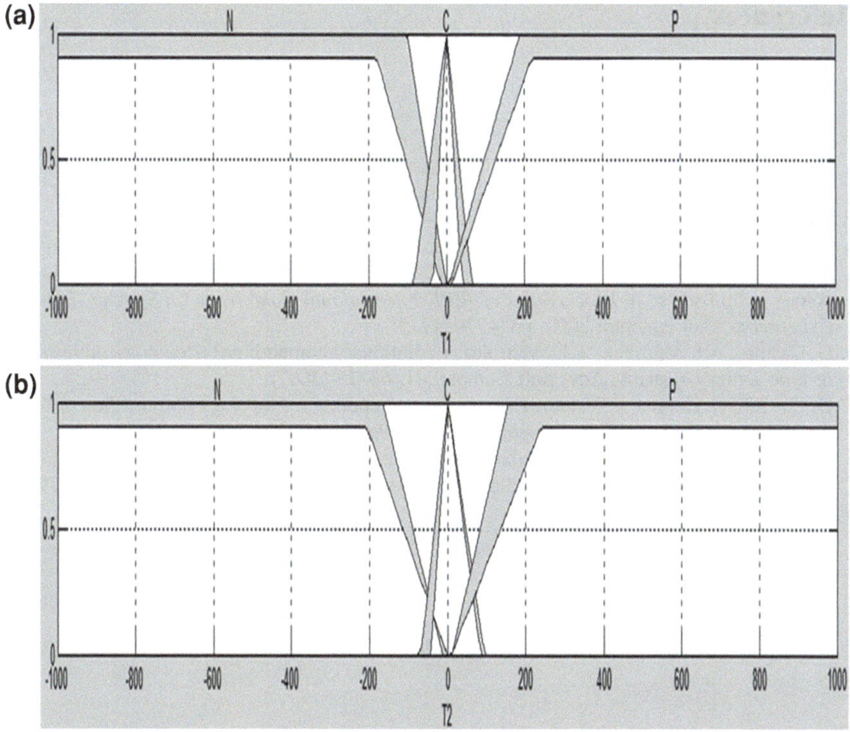

Fig. 8.16 Resulting output membership functions: **a** right and **b** left torque

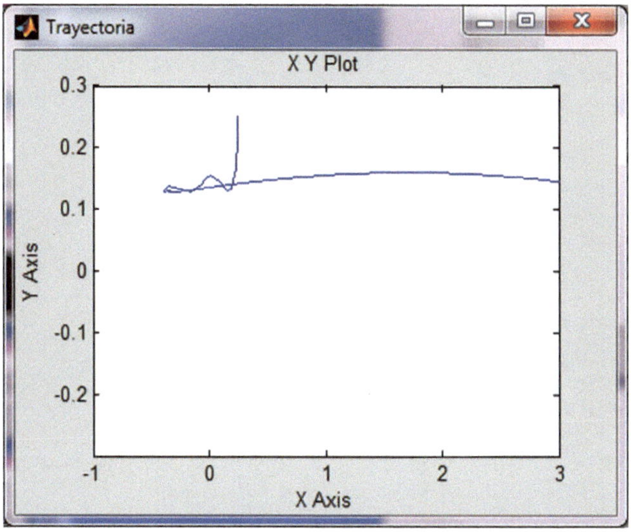

Fig. 8.17 Obtained trajectory when applying the chemical reaction algorithm

References

1. O. Castillo, R. Martinez-Marroquin, P. Melin, F. Valdez, J. Soria, Comparative study of bio-inspired algorithms applied to the optimization of type-1 and type-2 fuzzy controllers for an autonomous mobile robot. Inf. Sci. **192**(1), 19–38 (2012)
2. L. Astudillo, O. Castillo, L.T. Aguilar, R. Martinez, Hybrid control for an autonomous wheeled mobile robot under perturbed torques. Lecture Notes in Computer Science, vol. 4529 (2007), pp. 594–603
3. R. Martinez-Marroquin, O. Castillo, J. Soria, Parameter tuning of membership functions of a type-1 and type-2 fuzzy logic controller for an autonomous wheeled mobile robot using ant colony optimization, in *Proceedings of IEEE International Conference on Systems, Man and Cybernetics*, San Antonio, 2009, pp. 4770–4775
4. O. Castillo, A.I. Martinez, A.C. Martinez, Evolutionary computing for topology optimization of type-2 fuzzy systems. Adv. Soft Comput. **41**, 63–75 (2007)
5. O. Castillo, G. Huesca, F. Valdez, Evolutionary computing for topology optimization of type-2 fuzzy controllers. Stud. Soft Comput. **208**, 163–178 (2008)
6. R. Martinez, O. Castillo, L.T. Aguilar, Optimization of interval type-2 fuzzy logic controllers for a perturbed autonomous wheeled mobile robot using genetic algorithms. Inf. Sci. **179**(13), 2158–2174 (2009)

Chapter 9
Genetic Optimization of Interval Type-2 Fuzzy Systems for Hardware Implementation on FPGAs

This chapter proposes a method for the design of a Type-2 Fuzzy Logic Controller (FLC-T2) and a Type-1 Fuzzy Logic Controller (FLC-T1) using Genetic Algorithms. The two controllers were tested with different levels of uncertainty to Regulate Speed in a Direct Current Motor (ReSDCM). The controllers were synthesized in Very High Description Language (VHDL) code for a Field Programmable Gate Array (FPGA), using the Xilinx System Generator (XSG) of Xilinx ISE and Matlab-Simulink. Comparisons were made between the FLC-T1 versus FLC-T2 in VHDL code and also with a Proportional Integral Differential (PID) Controller, to ReSDCM. To evaluate the difference in performance of the three types of controllers, the t-student statistical test was used.

9.1 Introduction

Fuzzy logic systems are used successfully in many application areas, and these include control, classification, etc. [1, 2].

These systems based on rules incorporate linguistic variables, linguistic terms and fuzzy rules. The acquisition of these rules is not an easy task for the expert and is of vital importance in the operation of the controller.

The process of adjusting these linguistic terms and rules is usually done by trial and error, which implies a difficult task, and for this reason there have been methods proposed to optimize those elements that over time have taken importance, such as genetic algorithms [3–5].

There is a great interest in research and development of fuzzy systems, especially those based on type-2 fuzzy logic due to their advantages in the management of uncertainty, which is why there are great expectations regarding their use in control systems as a possible way to compensate for errors due to instrumentation systems, among others. However, these systems require large computing resources

O. Castillo and P. Melin, *Recent Advances in Interval Type-2 Fuzzy Systems*,
SpringerBriefs in Computational Intelligence, DOI: 10.1007/978-3-642-28956-9_9,

that difficultly can be provided by a personal computer, so that the investigation of different alternatives for their implementation is a topic of current research such as fuzzy control implemented in programmable logic devices.

9.2 Preliminaries

9.2.1 FPGA

An FPGA is a semiconductor device that contains in its interior components such as gates, multiplexers, etc. These are interconnected with each other, according to a given design. These devices use the VHDL programming language, which is an acronym that represents the combination of VHSIC (Very High Speed Integrated Circuit) and HDL (Hardware Description Language) [6].

The design of an FPGA implementation is done by specifying the logic function to develop, either by computer aided design (CAD) or through a hardware description language. Having defined the function to perform, the design is transferred to the FPGA. This process programs the configurable logic blocks (CLBs) to perform a specific function (there are thousands of configurable logic blocks in the FPGA). The configuration of these blocks and their interconnections flexibility are the reasons why the FPGA can implement very complex designs. The interconnections enable connecting the CLBs. Finally, it has configuration memory cells (CMC, Configuration Memory Cell) distributed throughout the chip, which store all the information necessary for programming the mentioned programmable elements. These cells usually consist of a RAM configuration and are initialized in the process of loading the configuration. The programmable elements of an FPGA are:

1. Configurable Logic Blocks (CLBs)
2. In/Out Blocks (IOBs)
3. Programmable Interconnection
   - By fuse technology and be of OTP.
   - By antifuses or by type SRAM cells.

Figure 9.1 shows these basic elements.

Depending on the manufacturer we can find different solutions. FPGAs currently available on the market, depending on the structure adopted by the logical blocks that are defined, can be classified as belonging to four major families Xilinx, Orca, Actel and Altera [6]. FPGAs currently available on the market, depending on the structure adopted by the logical blocks that are defined, can be classified as belonging to four major families shown in Fig. 9.2.

Figure 9.3 shows the Spartan chip Basic elements.

FPGAs are good platforms for fast prototyping of digital hardware. FPGAs can be programmed into the system, without shutting down the system.

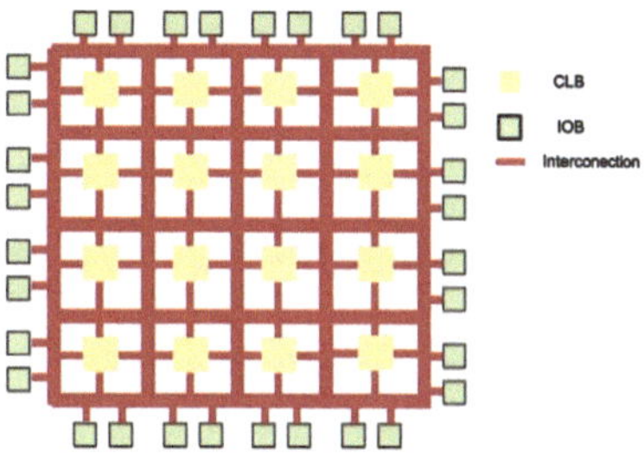

Fig. 9.1 FPGA basic elements

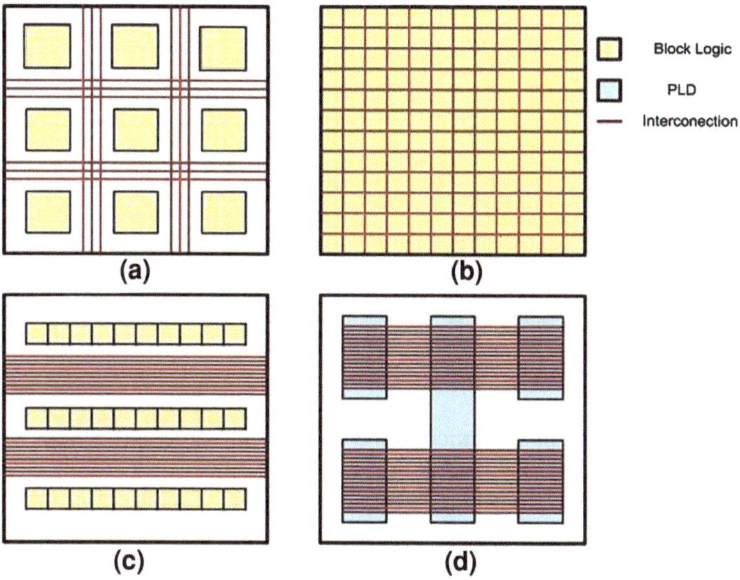

Fig. 9.2 Block logic **a** symmetrical array (Xilinx), **b** sea of gates (ORCA), **c** row based (ACTEL), and **d** hierarchical PLD (Altera and Xilinx)

This functionality allows modification and tuning of rules and-or fuzzifiers to achieve better control performance. The implementation of type-2 fuzzy systems onto FPGAs has been investigated by several researchers [6], but it is still a subject of current research.

The FPGAs can be used to implement specific architectures to accelerate a particular algorithm. Applications that require a great number of simple operations are suitable for implementation on FPGAs. A processing element can be designed to perform this operation and several instances of it can be used to perform parallel processing [6].

Any hardware implementation of an electronic system requires a complex methodology to test and validate every stage in the design process to guarantee its

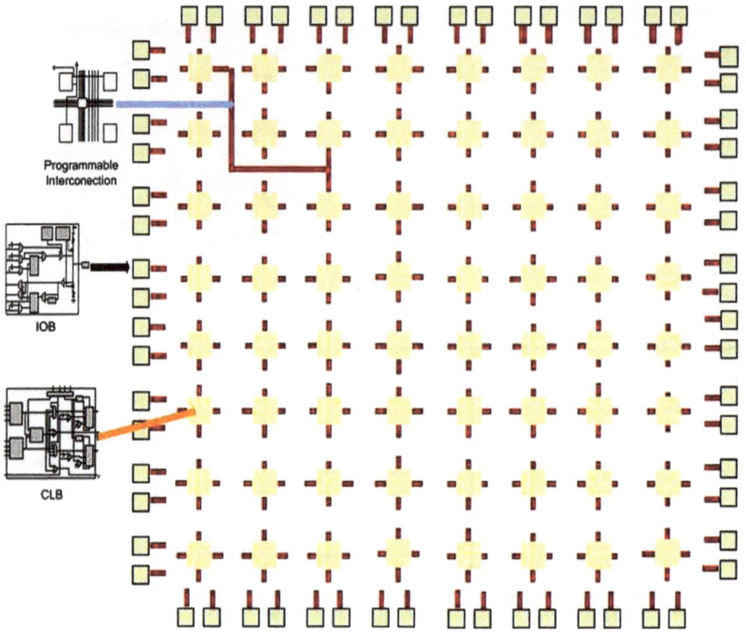

Fig. 9.3 CHIP Spartan of Xilinx basic elements

correct functionality; this is particularly true when the designer decides to use a HDL to make a design.

These systems based on rules incorporate linguistic variables, linguistic terms and fuzzy rules. The acquisition of these rules is not an easy task for the expert and is of vital importance in the operation of the controller. There are methods to optimize those elements, such as the genetic algorithm (GA) [4].

9.2.2 Genetic Algorithms

A Genetic Algorithm (GA) [4, 7] is a stochastic optimization algorithm inspired by the natural theory of evolution. From a principle proposed by Holland [8], the GA has been used successfully to manage wide variety problems such as control, search, etc. [9].

GAs are search algorithms based on the mechanics of natural selection. The combination of survival of the fittest among string structures with a structured yet randomized information exchange to form a search algorithm with some of the innovative flair of human search provides a good optimization method. In every generation, a new set of artificial creatures is created using bits and pieces of the fittest of the old individuals; an occasional new part is also tried for good measure.

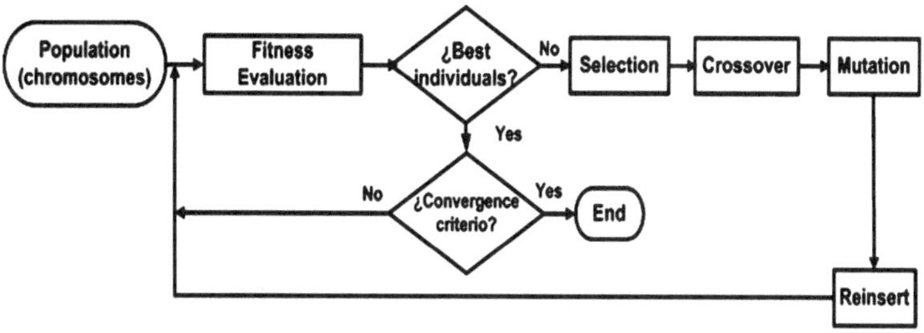

Fig. 9.4 GA cycle

The GA is inspired by the mechanism of natural selection where stronger individuals are likely to be the winners in a competing environment. Here the GA uses a direct analogy of such natural evolution. Through the genetic evolution method, an optimal solution can be found and represented by the final winner of the genetic game. Figure 9.4 shows the corresponding GA cycle.

The GA has applications in a wide variety of fields to develop solutions to complex problems, including optimization of fuzzy systems, offering them learning and adaptation capabilities in this case, they are commonly called genetic fuzzy systems or fuzzy system hybrids.

A GA allows a population composed of many individuals to evolve under specified selection rules to a state that maximizes the "fitness" (i.e. minimizes the cost function).

9.2.3 Type-1 Fuzzy Inference System

Type-1 Fuzzy inference systems (FIS-T1) have recently been used more frequently, because they tolerate imprecise information and can be used to model nonlinear functions of arbitrary complexity. A fuzzy inference system (FIS) consists of three stages: Fuzzification, Inference and Defuzzification [7]. In Fig. 9.5 the fuzzy system information processing is shown.

We describe below these stages.

1. Fuzzification: Is the interpretation of input values (numeric) by the fuzzy system, and the obtained output are fuzzy values. Let $x_{\in}X$ be a linguistic variable and $T_i(x)$ a fuzzy set associated with a linguistic value T_i. The translation of a numeric value x corresponds to a linguistic value associated with a degree of membership, $x \rightarrow \mu_{Ti}(x)$, and this is known as Fuzzification. The membership degree $\mu_{Ti}(x)$ represents a value of membership to a fuzzy set [7].
2. Inference: Is basically like the brain of the system, here the rules of the if–then form that describe this behavior are used [7]. For example: *If* x_1 *is* A_1 *and* … *and* x_n *is* A_n *Then* y *is* B, where x_1, …, x_n are the inputs, A_1, …, A_n, B are linguistic terms and $y$is the output.

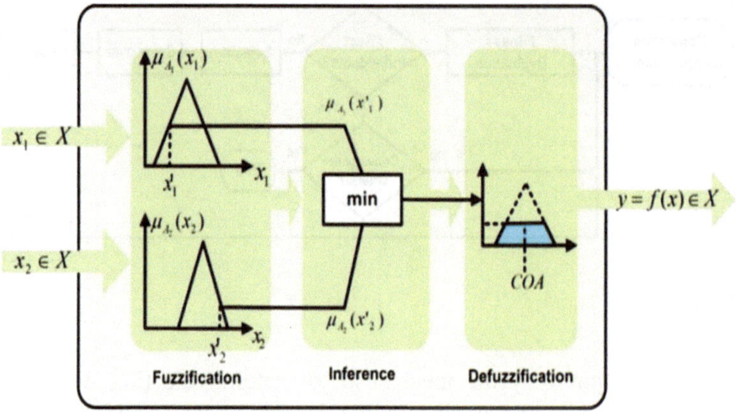

Fig. 9.5 Type-1 fuzzy systems

3. Defuzzification: Consists in obtaining a numeric value for the output. This stage basically selects a point that is the most representative of the action to perform [7]. There are several methods to calculate the Defuzzification, such as the Center of Height (COA), Center of Gravity (COG), etc. The COG is shown in (9.1).

$$y = \frac{\sum_{i=1}^{N} h_i \mu(i)}{\sum_{i=1}^{N} \mu(i)} \tag{9.1}$$

where h_1 is the maximum height of the consequent from rule i to rule N [7].

9.2.4 Type-2 Fuzzy Inference Systems

An FIS-T1 uses exact membership functions, while type-2 fuzzy inference systems (FIS-T2) are described by membership functions with uncertainty.

Uncertainty is an attribute of information, which may be incomplete, inaccurate, vague, weak, contradictory, and so on [7].

The FIS-T2 consists of four stages: Fuzzification, Inference, Type Reduction and Defuzzification. We describe below these stages. The fuzzification maps a numeric value into a type-2 fuzzy set A_x in X. A_x is a singleton fuzzy set if $\mu_{\mathrm{A}x} = 1/1$ for $x = x'$ and $\mu_{\mathrm{A}x} = 1/0$ for all others x'. The Inference stage consists of two blocks, the rules and the inference engine, it works the same way as for type-1 fuzzy systems, and except the antecedents fuzzy sets and the consequent are represented by type-2 fuzzy sets. The process consists of combining the rules and map the input to the output (type-2 fuzzy sets), using the Join and Meet operations. The Type Reductor is used to convert all type-2 fuzzy sets to type-1 fuzzy intervals

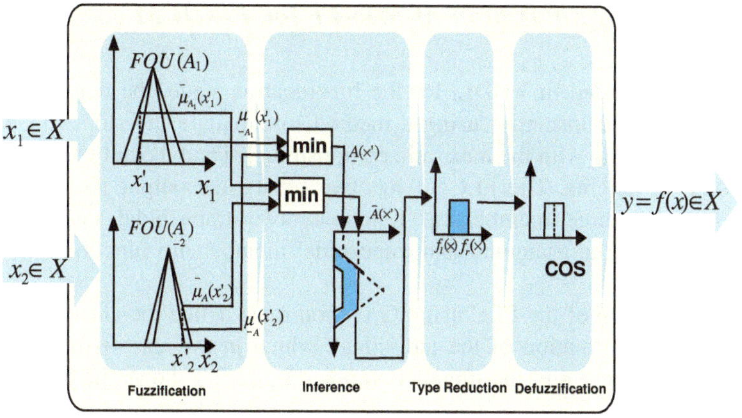

Fig. 9.6 FIS-T2 information processing

on the output. There are several methods to calculate the reduced set, such as joint center, center of sums (COS), height, among others [7]. The Defuzzification stage consists in obtaining a numeric value for the output. Using the COS type reductor, the defuzzification is an average value since the range is given by $[y_l, y_r]$.

In Fig. 9.6, the FIS-T2 information processing is shown.

An FIS-T2 can be implemented on a general purpose computer, or by a specific use of a microelectronics realization such as the FPGA.

This paper proposes a method for genetic optimization of the triangular and trapezoidal membership functions (MF) of a type-2 fuzzy logic controller (FLC-T2) for hardware applications such as the FPGA. This method involves taking only certain points of the membership functions in order to give greater efficiency to the algorithm. The GA has been tested in a FLC-T2 and FLC-T1 to regulate the speed of a direct current motor using the Matlab platform and Xilinx System Generator (XSG) [6]. Comparisons were made between the FLC-T1 versus FLC-T2 in VHDL code and FLC-T2 versus PID Controller, to regulate the velocity of a DC motor, to evaluate the difference in performance of three types of controllers, using the t-student statistical test.

9.3 Genetic Optimization of Type-1 and Type-2 Membership Functions for the Regulation of Speed of a DC Motor

We optimized the type-1 membership functions (MF-T1) and type-2 membership functions (MF-T2) for ReSDCM, and below we explain the process of optimization for each case.

9.3.1 Genetic Optimization of MF-T1 for ResDCM

The FLC-T1 is coded in VHDL, for the fuzzification stage, the degree of membership is obtained instantly, using a method to calculate the slopes [5, 7], the inference is working with the max–min composition [4] and the defuzification with the method of heights. The FLC-T1 has two inputs one output. Each input and output contains three membership functions, two trapezoidal one triangular. Figure 9.7 shows the triangular and trapezoidal membership functions (MF) that are used.

For optimization of the FLC using GAs, you must define the chromosome that represents the information of the individual, which in this case is related to the universe of discourse and the linguistic terms. Figure 9.8 shows the chromosome of the GA.

In Table 9.1, we show the boundary parameters of the chromosome.

Figure 9.9 shows the input of the FLC with fixed and variable parameters. Each input and output has a size of 8 bits.

The blue points are fixed, the red dots are the parameter a_2, the green dots are fixed (b_1) and the yellow dots are the parameter a_1.

In Fig. 9.10 the parameter ranges of the membership functions are shown, where a_1 and a_2 correspond to the membership functions 1 and 3, respectively, for FLC inputs and output.

In Fig. 9.11 the VHDL code for the triangular FM-T1 for the fuzzification stage is shown.

In Fig. 9.12 the VHDL code for the trapezoidal FM-T1 for the fuzzification stage is shown.

In Fig. 9.13 the block diagram of FLC-T1 in VHDL code for Matlab-XSG is shown.

In Fig. 9.13 the first blue block represents the stage of fuzzification that is connected to the stage of inference and in turn to the defuzzification stage. The FLC-T1 has a single output and as inputs: reset, error and change of error, 22 parameters of the FM-T1 for the inputs(error and change of error) and 3 parameters of the FM-T1 for output(signal of control).

9.3.2 Genetic Optimization of MF-T2 for ReSDCM

We implemented the FLC-T2 using the average method [10] in the FPGA. The FLC-T2 is coded in VHDL, the FLC for the fuzzification stage, is able to instantly calculate the degree of membership, using a method to calculate the slopes [6], the inference is working with the max–min composition and the defuziffication with the method of heights [7].

Figure 9.14 shows the block diagram of the average FLC-T2 for ReSDCM, the system inputs are the error (x_1), and change of error (x_2). The system has only one output (y).

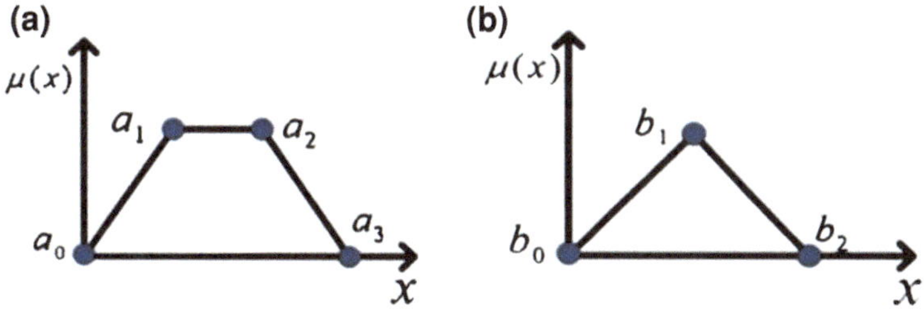

Fig. 9.7 Parameters of the membership functions: **a** MF trapezoidal, **b** MF triangular

Fig. 9.8 GA chromosome for the type-1 fuzzy system

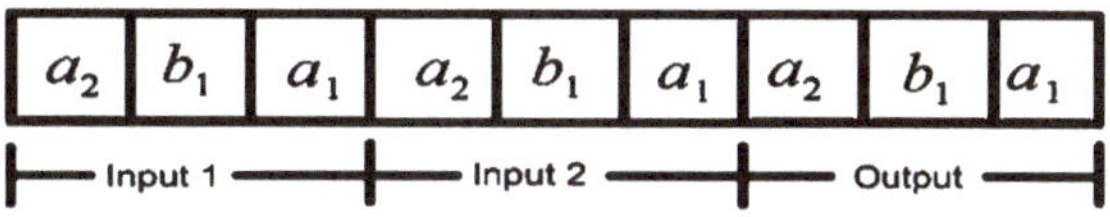

Table 9.1 Boundary parameters of the chromosome

	Input 1	Input 2	Output
Parameters	$0 < a_2 < 128$ $b_1 = 128$ $128 < a_1 < 255$	$0 < a_2 < 128$ $b_1 = 128$ $128 < a_1 < 255$	$0 < a_2 < 128$ $b_1 = 128$ $128 < a_1 < 255$

Fig. 9.9 Points of the membership functions input and output

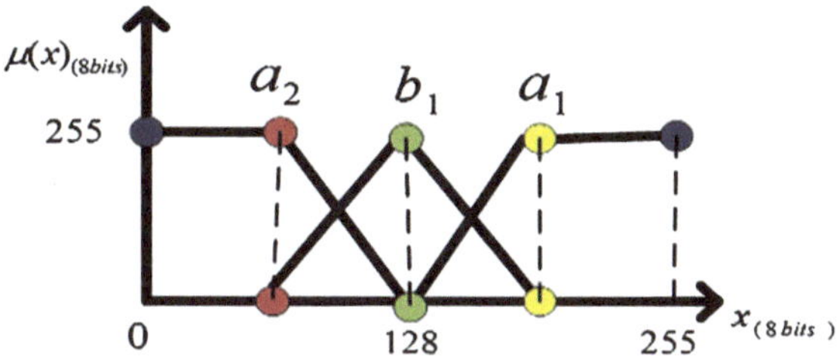

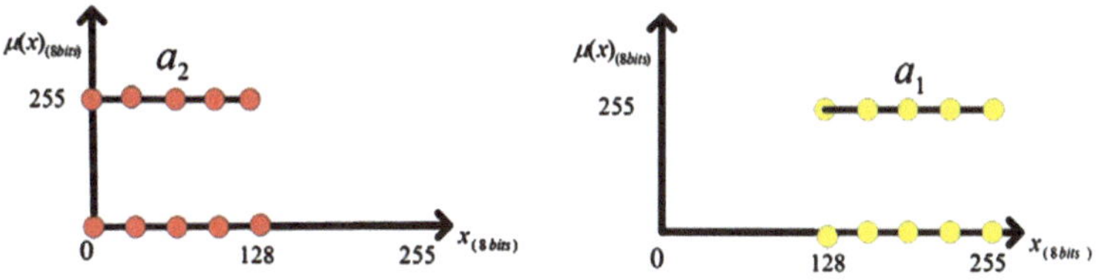

Fig. 9.10 Range of parameters membership functions

```
function trian (x: std_logic_vector; y:triangular; m: pendiente)return std_logic_vector is
         variable sal: std_logic_vector(n downto 1):=(others =>'0');
         variable sal1: std_logic_vector(n*2 downto 1):=(others =>'0');

  begin
    if x<=y(1) then sal:="00000000";
      elsif x> y(1) and x<y(2)
        then
        sal1:=m(1)*(x-y(1));
        sal := sal1(8 downto 1);
      elsif x= y(2) then sal:=tope;
      elsif x> y(2) and x< y(3)
        then
        sal1:= m(2)*(y(3)-x);
        sal := sal1(8 downto 1);
      else sal:="00000000";
    end if;
  return sal;
  end function trian;
```

Fig. 9.11 VHDL code for triangular FM-T1

```
function trape (x: std_logic_vector; y:trapezoidal; m: pendiente)return std_log
         variable sal: std_logic_vector(n downto 1):=(others =>'0');
         variable sal1: std_logic_vector(n*2 downto 1):=(others =>'0');

  begin
    if (x<=y(1)) then
      if (y(1) = y(2))then
        sal := tope;
      else
        sal:="00000000";
      end if;
      elsif (x> y(1) and x<y(2)) then
        sal1:=m(1)*(x-y(1));
        sal := sal1(8 downto 1);
      elsif x>= y(2) and x<= y(3) then
        sal:=tope;
      elsif x> y(3) and x<= y(4) then
        if (y(3) = y(4))then
        sal:= tope;
      else
        sal1:= m(2)*(y(4)-x);
        sal := sal1(8 downto 1);
        end if;
      else
        sal:="00000000";
    end if;
  return sal;
  end function trape;
```

Fig. 9.12 VHDL code for trapezoidal FM-T1

The FLC-T2 has two inputs and one output, each input and output contains three membership functions, two trapezoidal one triangular. Figure 9.15 shows the triangular and trapezoidal membership functions (MF) that are used.

For the optimization of the FLC-T2 using GAs, you must define the chromosome that represents the information of the individual, which in this case is related

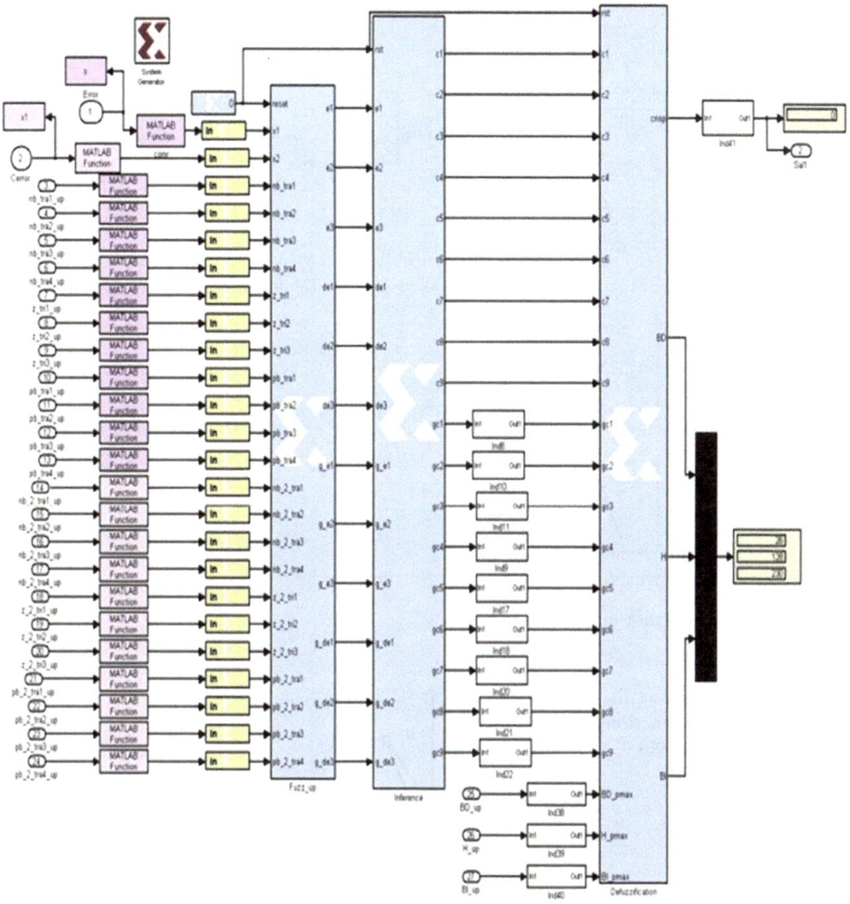

Fig. 9.13 Block diagram of the FLC-T1 in VHDL code for Matlab-XSG

to the universe of discourse and the linguistic terms. Figure 9.16 shows the chromosome used for the GA.

In Table 9.2, we show the boundary parameters of the chromosome for this case.

Figure 9.17 shows the input of the FLC-T2 with fixed and variable parameters. Each input and output has a size of 8 bits.

The blue points are fixed, the red dots represent the parameter a_2, the green dots are fixed (b_1) and the yellow dots represent the parameter a_1.

In Fig. 9.18 the block diagram of the average FLC-T2 in VHDL code for Matlab-XSG is shown.

Figure 9.18 corresponds to the FLC-T2, the top part simulates the FLC-T1 and the bottom another FLC-T1, the result is the average of these two fuzzy systems.

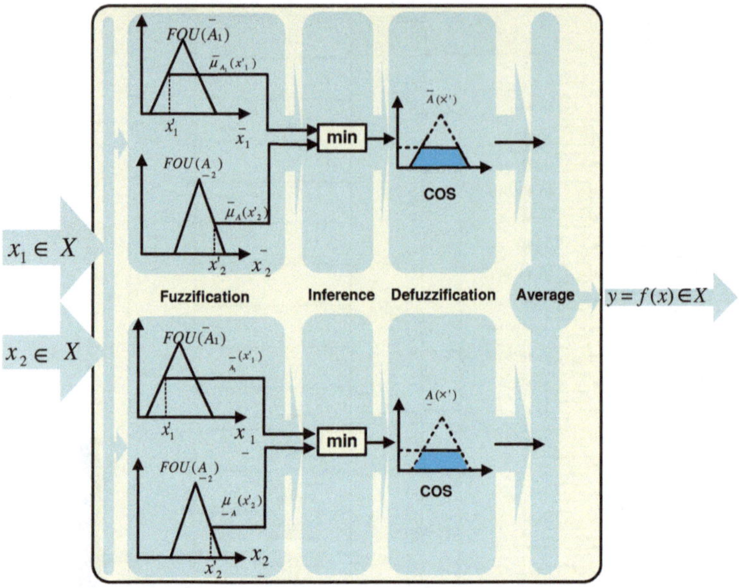

Fig. 9.14 Block diagram in XSG of FLC

Fig. 9.15 Parameters of the type-2 membership functions. **a** MF trapezoidal, **b** MF triangular

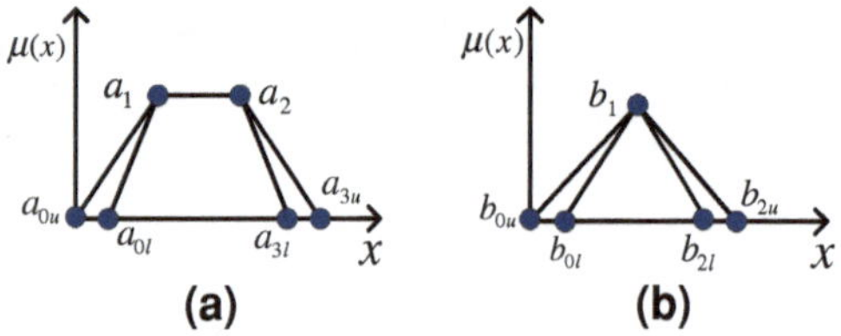

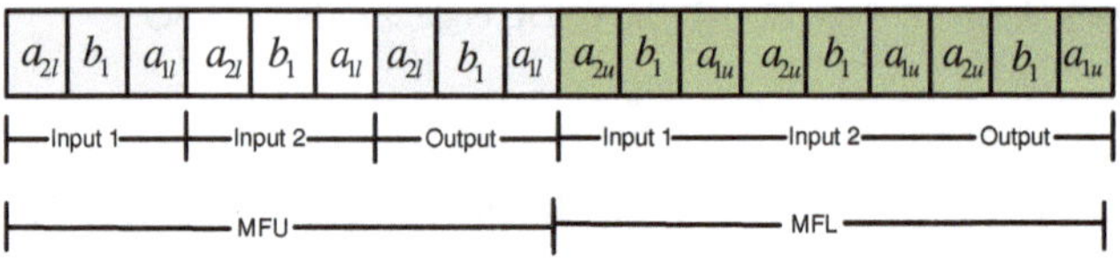

Fig. 9.16 GA chromosome for the type-2 fuzzy system

Each FLC is independent, but they share the same inputs with different FMs, this is used to simulate the uncertainty of the real the system.

The GA is of multiobjective type [4] for both the FLC-T1 and FLC-T2, which means that to determine the best individual, three evaluations are performed:

(a) Minimum overshoot

$$o_1 = \max(y(t)) - r \qquad (9.2)$$

Table 9.2 Boundary parameters of the chromosome type-2

	Input 1	Input 2	Output
Parameters U (P_U)	$0 < a_{2U} < 128$ $b_{1U} = 128$ $128 < a_{1U} < 255$	$0 < a_{2U} < 128$ $b_{1U} = 128$ $128 < a_{1U} < 255$	$0 < a_{2U} < 128$ $b_{1U} = 128$ $128 < a_{1U} < 255$
Parameters L (P_L)	P_U && $a_{2U} > a_{2L}$ $b_{1L} = 128$ P_U && $a_{2L} > a_{2U}$	P_U && $a_{2U} > a_{2L}$ $b_{1L} = 128$ P_U && $a_{2L} > a_{2U}$	P_U && $a_{2U} > a_{2L}$ $b_{1L} = 128$ P_U && $a_{2L} > a_{2U}$

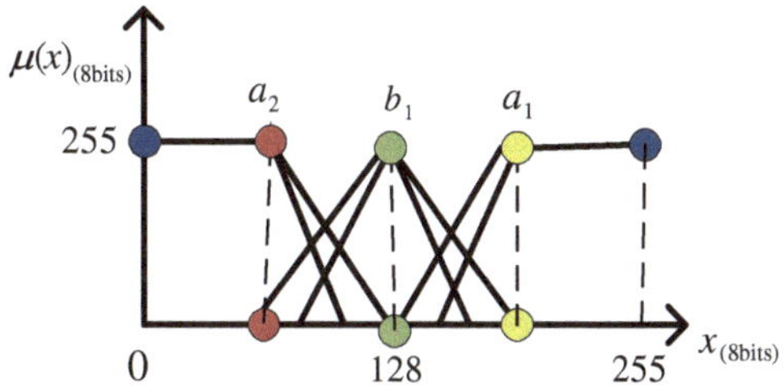

Fig. 9.17 Points of the input and output of type-2 membership functions input and output

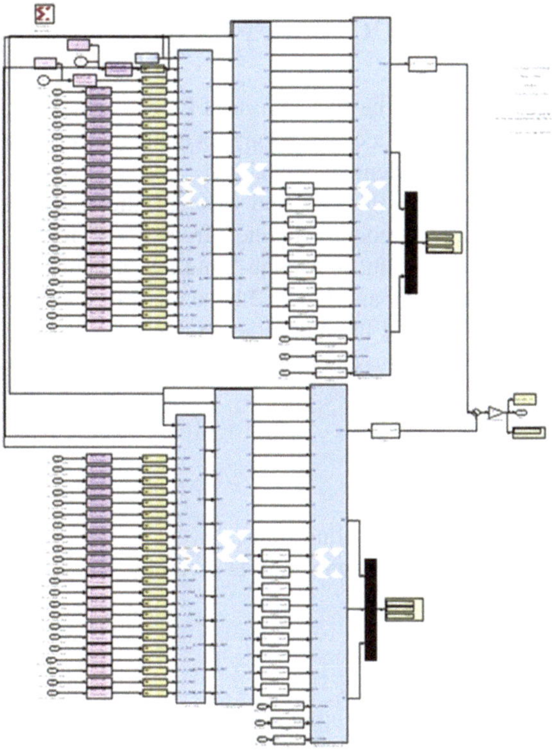

Fig. 9.18 Block diagram of FLC-T2 in VHDL for Matlab-XSG

(b) Minimum undershoot

$$o_2 = |\min(y(t)) - r| \tag{9.3}$$

(c) Minimum output steady state error (sse)

$$sse = \sum_{t=201}^{1000} y(t) - r \tag{9.4}$$

Where $y(t)$ is the output of the system and r is reference.

The FLC linguistic terms were optimized with the GA, but the fuzzy rules are not changed. The process of the GA is described below: Generate the initial population, Fitness Evaluation (o_1, o_2, sse), Selection, Crossover, Mutation, Reinsert and Simulation using the XSG plataform in Matlab- Simulink [5]. Figure 9.19 shows the GA process for the FLC-T1 and FLC-T2.

9.4 Test and Results of the FLC-T1 and FLC-T2 for ReSDCM in FPGAs

In this section the FLC-T1 and FLC-T2 are analyzed; each was given a level of uncertainty and a comparison was made between them. The results were evaluated using the t-student statistical rest.

To test the FLC-T2 and FLC-T1 the speed control was simulated using a mathematical model (obtained from a DC motor Pittman GM9236S025-R1 of 12 V) of the plant in Matlab-Simulink, as shown in Fig. 9.20.

The FLC-T1 and FLC-T2 have the following inputs, error ($e(t)$) and change of error ($e'(t)$), and the output is the control signal ($y(t)$).

The inputs are calculated as follows:

$$e(t) = r(t) - y(t) \tag{9.5}$$

$$e'(t) = e(t) - e(t-1) \tag{9.6}$$

where t is the sampling time.

The reference signal $r(t)$, is given by:

$$r(t) = \begin{cases} 15 & t > 0 \\ 0 & t \leq 0 \end{cases} \tag{9.7}$$

Each input and output of the FIS-T2 and FIS-T1 has three linguistic terms. For the linguistic variables of error and change of error, the terms are {NB, Z, PB}, in this case NB is Negative Big, Z is Zero and PB is Positive Big. For the linguistic variable control signal, the terms are {BD, H, BI}, in this case BD is Big

Fig. 9.19 Optimization GA

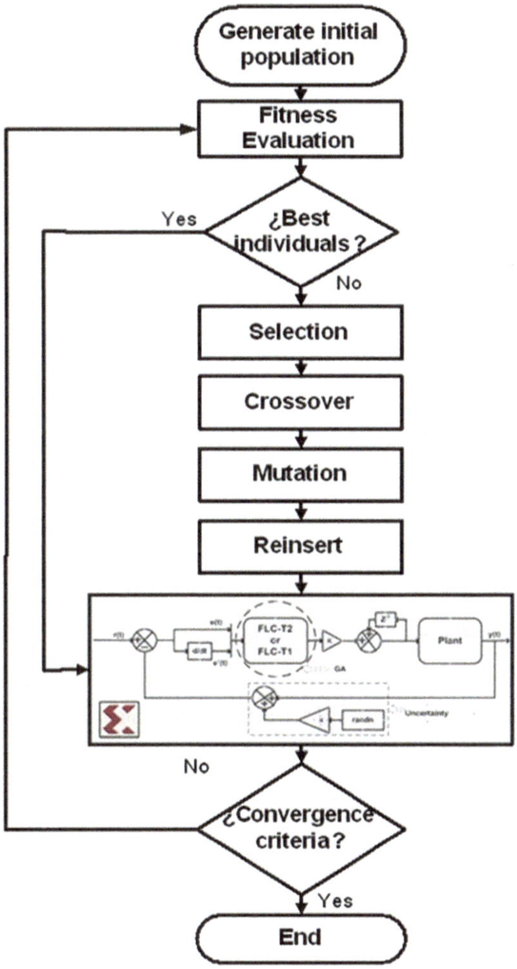

Decrement, H is Hold and BI is Big Increment. Table 9.3 shows the rule matrix of both the FLC-T1 and FLC-T2.

A series of experiments for the FLC-T2 were performed and are listed on Table 9.4.

In experiment No. 18 the best FLC-T2 was found because this has the lower error value. Below are the FLC-T2 characteristics for experiment 18.

Figure 9.21 shows the FM-T2 of the error input due to the behavior of the GA for the best FLC-T2.

Figure 9.22 shows the FM-T2 of the change of error input for the FLC-T2.

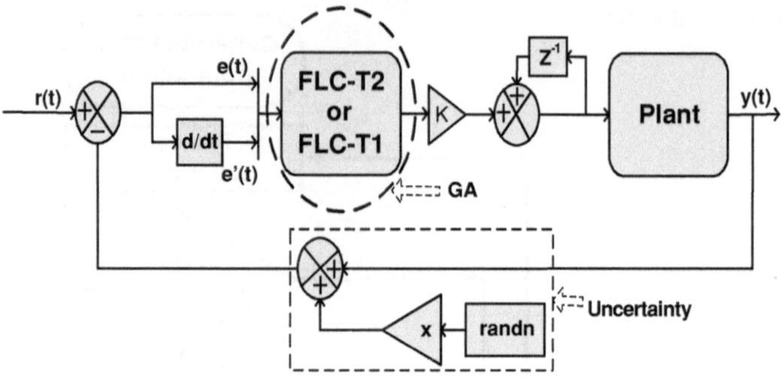

Fig. 9.20 Model of FLC-T2 and FLC-T1

Table 9.3 Rule matrix

e \ ce	NB	Z	PB
NB	BD	BD	BD
Z	H	H	H
PB	BI	BI	BI

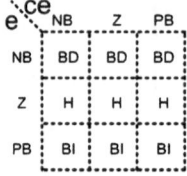

Table 9.4 FLC-T2 results for different experiments

No.	Generations	Crossover (XOVSP)	Selection (SUS)	Mutation	Error	Time (s)
1	30	0.75	0.75	0.1	01282	16.416
2	30	0.75	0.75	0.1	0.1282	16.778
3	30	0.75	0.75	0.1	0.1282	16.252
4	16	0.8	0.9	0.1	0.0603	10.157
5	25	0.7	0.75	0.1	0.0785	19.086
6	20	0.7	0.75	0.1	0.0785	16.879
7	11	0.5	0.75	0.1	0.1172	11.468
8	40	0.7	0.75	0.1	0.0785	20.858
9	24	0.75	0.75	0.1	0.0603	19.571
10	11	0.55	0.75	0.05	0.1689	21.694
11	40	0.69	0.75	0.1	0.75	22.217
12	30	0.75	0.6	0.05	0.1897	38.8286
13	11	0.75	0.75	0.05	0.1897	15.347
14	11	0.75	0.85	0.05	0.1897	29.4589
15	24	0.75	0.85	0.1	0.0603	34.0213
16	17	0.85	0.85	0.2	0.0832	17.4676
17	18	0.85	0.85	0.13	0.1198	19.7046
18	18	0.85	0.8	0.09	0.0345	15.6772
19	17	0.8	0.8	0.1	0.0603	28.3152
20	11	0.75	0.6	0.1	0.1172	14.0033
21	30	0.69	0.75	0.1	0.078	20.2698
22	30	069	075	01	00781	200744

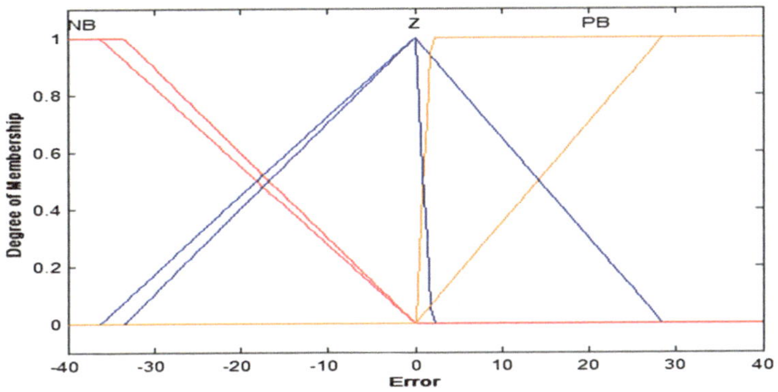

Fig. 9.21 Behavior of GA for FLC-T2 for input $e(t)$

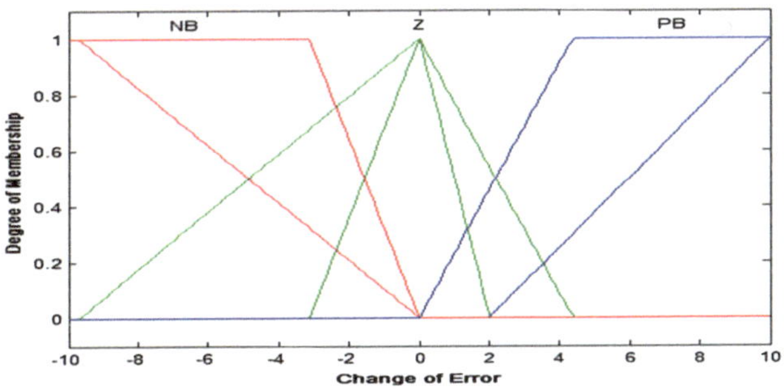

Fig. 9.22 Behavior of GA for FLC-T2 for input $e'(t)$

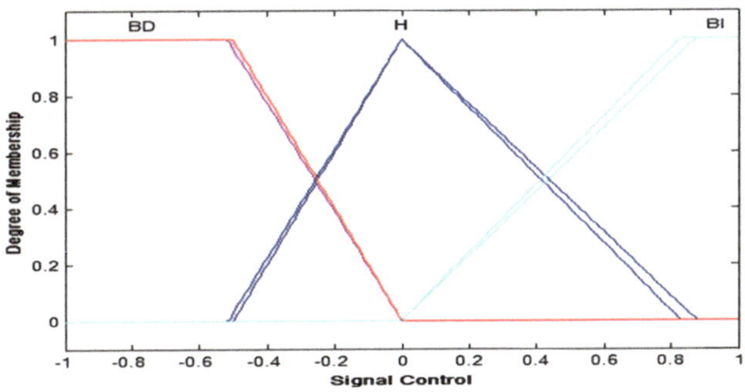

Fig. 9.23 Behavior of GA for FLC-T2 for output $y(t)$

Figure 9.23 shows the FM-T2 of the output due to the behavior of the GA for the FLC-T2.

Figure 9.24 shows the motor velocity due to the behavior of the GA for FLC-T1 versus FLC-T2.

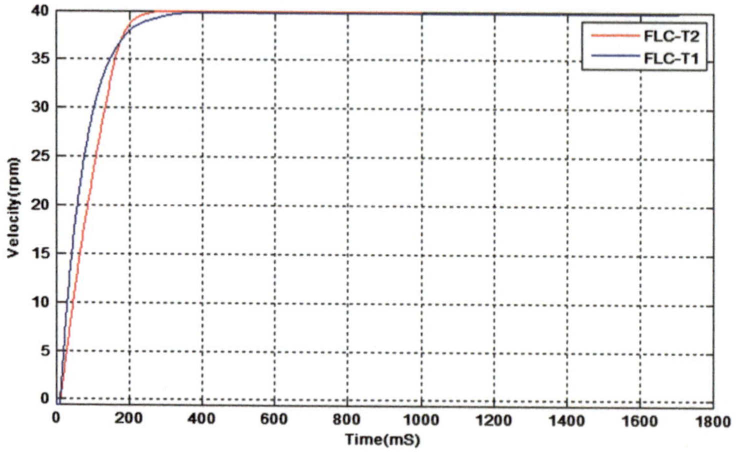

Fig. 9.24 Behavior of GA for FLC-T1 versus FLC-T2 for velocity motor

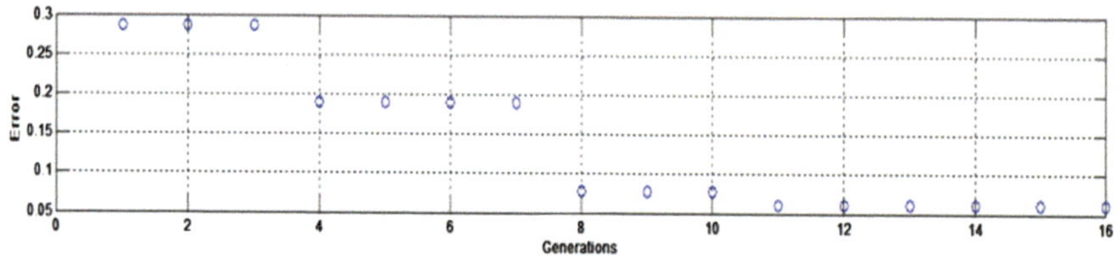

Fig. 9.25 Behavior of GA for FLC-T2 for error convergence

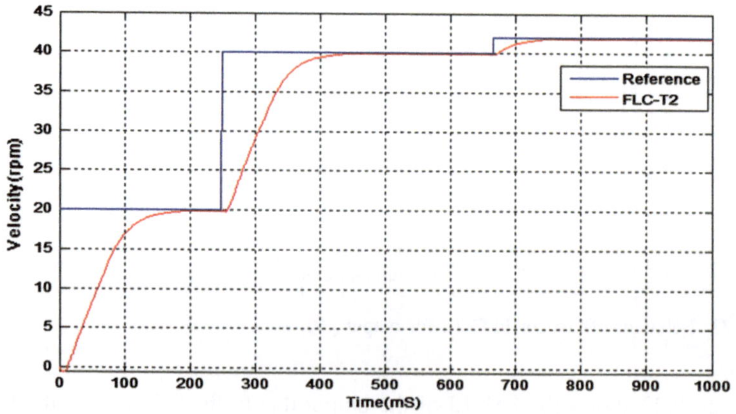

Fig. 9.26 Different motor velocities for the FLC-T2

Table 9.5 FLC-T1, FLC-T2 versus PID results for ReSDCM

No.	FLC	Uncertainty level factor	Error
1	T2	0	0.0120
	T1	0	0.1497
	PID	0	1.42e−4
2	T2	0.001	0.0456
	T1	0.001	0.1497
	PID	0.001	0.0014
3	T2	0.005	0.0456
	T1	0.005	0.1492
	PID	0.005	0.0070
4	T2	0.008	0.0456
	T1	0.008	0.1455
	PID	0.008	0.0112
5	T2	0.05	0.0255
	T1	0.05	0.0975
	PID	0.05	0.0700
6	T2	0.08	0.0014
	T1	0.08	0.0699
	PID	0.08	0.112
7	T2	0.1	0.0053
	T1	0.1	0.0536
	PID	0.1	0.1400
8	T2	0.2	0.0585
	T1	0.2	0.0354
	PID	0.2	0.2799
9	T2	0.3	0.0014
	T1	0.3	0.0551
	PID	0.3	0.4199
10	T2	0.4	0.0255
	T1	0.4	0.0750
	PID	0.4	0.5598
11	T2	0.5	0.0120
	T1	0.5	0.0700
	PID	0.5	0.6998
12	T2	0.6	0.0893
	T1	0.6	0.0978
	PID	0.6	0.8398
13	T2	0.7	0.0389
	T1	0.7	0.1044
	PID	0.7	0.9797
14	T2	0.8	0.1095
	T1	0.8	0.1242
	PID	0.8	1.1197
15	T2	0.9	0.1767

(continued)

Table 9.5 (continued)

No.	FLC	Uncertainty level factor	Error
	T1	0.9	0.1439
	PID	0.9	1.2597
16	T2	1	0.2372
	T1	1	0.1689
	PID	1	1.3996

Table 9.6 FLC-T1, FLC-T2 versus PID results for velocity regulation in a dc motor

Controllers comparison	t-student
FLC-T1 versus PID	3.13
FLC-T2 versus PID	3.5
FLC-T2 versus FLC-T1	2.41

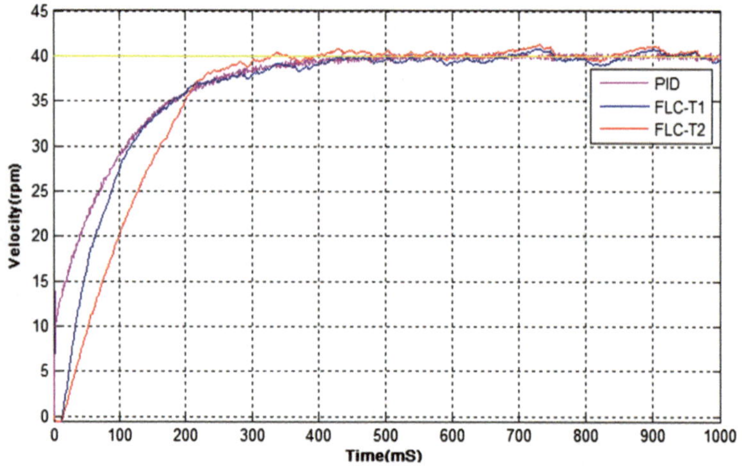

Fig. 9.27 Behavior of FLC-T2 comparison with FLC-T1 and PID controllers for velocity motor with uncertainty level (x = 1)

Figure 9.25 shows the convergence error due to the behavior of the GA for the FLC-T2.

Figure 9.26 shows the different motor velocities for the FLC-T2.

In Table 9.5, we show the comparison between the FLC-T1, FLC-T2 versus the PID controller for different levels of uncertainty. We note that the FLC-T2 is better at different levels of uncertainty (noise), while the noise free FLC-T1 has similar behavior to the FLC-T2, while in this case the PID is better.

We analyze statistically the performance of the three controllers using the t-student test. Table 9.6 shows the statistical results of the three controllers.

As shown in Table 9.6, the FLC-T2 has on average a better performance compared with the FLC-T1 and PID, with a degree of confidence of more than 95 percent.

Figure 9.27 shows the velocity of the FLC-T2 in comparison with the FLC-T1 and PID controllers with a particular level of uncertainty ($x = 1$).

As shown in Fig. 9.27, it is very difficult to determine which controller has better performance, for that reason we decided to use the test t-student statistical test shown in Table 9.6 which tells us that the FLC-T2 is better when compared to the FLC-T1 and PID controllers, for this appplication.

9.5 Summary

We described the genetic optimization of FLC-T1 and FLC-T2 for the ReSDCM, where three triangular and trapezoidal membership functions for the two inputs and one output are used in the optimization. The GA only optimizes parameters of the membership functions, but the rules are not optimized because we are interested in the speed of the algorithm. The objective function of the GA considers three characteristics: overshoot, undershoot and steady state error, so that makes it a multiobjective GA.

The FLC-T1 and FLC-T2 are encoded on VHDL code for implementation in the FPGA.

The best FLC-T2 was obtained in 18 generations with 85% crossover (single point crossover) and 80% selection (universal selection) and 9% Mutation rate, with an error of convergence of 0.0345, in a time of 15.67772 min with a speed of 40 rpm.

The PID controller tuning was performed with Ziegler-Nichols method and the obtained values of the constants are $k_p = 0.5$, $k_i = 0.2$ and $k_d = 0.025$.

Comparisons were made between the FLC-T1 versus FLC-T2 in VHDL code and FLC-T2 versus PID Controller, for ReSDCM, to evaluate the difference in performance of the three types of controllers, using the t-student statistical test, giving better results for the FLC-T2. Matlab-Simulink and XSG were used to perform the simulations in all cases.

References

1. M.O. Al-Jaafreh, A.A. Al-Jumaily, Training type-2 fuzzy system by particle swarm optimization, in *IEEE Congress on Evolutionary Computation, CEC 2007*, Singapore, 2007, pp. 3442–3446
2. L. Astudillo, O. Castillo, L.T. Aguilar, R. Martinez, Hybrid control for an autonomous wheeled mobile robot under perturbed torques. Lecture Notes in Computer Science, vol. 4529 (2007), pp. 594–603

3. O. Castillo, P. Melin, *Soft computing for control of non-linear dynamical systems* (Springer, Heidelberg, 2001)
4. O. Castillo, P. Melin, *Soft computing and fractal theory for intelligent manufacturing* (Springer, Heidelberg, 2003)
5. O. Castillo, A.I. Martinez, A.C. Martinez, Evolutionary computing for topology optimization of type-2 fuzzy systems. Adv. Soft Comput. **41**, 63–75 (2007)
6. R. Sepulveda, O. Montiel, G. Lizarraga, O. Castillo, Modeling and simulation of the defuzzification stage of a type-2 fuzzy controller using the Xilinx system generator and Simulink. Stud. Comput. Intell. **257**, 309–325 (2009)
7. O. Castillo, P. Melin, *Type-2 fuzzy logic: theory and applications* (Springer, Heidelberg, 2008)
8. N.S. Bajestani, A. Zare, Application of optimized type-2 fuzzy time series to forecast Taiwan stock index, in *Second International Conference on Computer, Control and Communication*, 2009, pp. 275–280
9. O. Castillo, G. Huesca, F. Valdez, Evolutionary computing for topology optimization of type-2 fuzzy controllers. Stud. Fuzziness Soft. Comput. **208**, 163–178 (2008)
10. T.W. Chua, W.W. Tan, Genetically evolved fuzzy rule-based classifiers and application to automotive classification. Lect. Notes in Computer Science, vol. 5361 (2008), pp. 101–110

Chapter 10
General Overview of the Area and Future Trends

In this chapter a general overview of the area of type-2 fuzzy system optimization is presented. Also, possible future trends that we can envision based on the review of this area are presented. It has been well-known for a long time, that designing fuzzy systems is a difficult task, and this is especially true in the case of type-2 fuzzy systems. The use of GAs, ACO and PSO in designing type-1 fuzzy systems has become a standard practice for automatically designing this sort of systems. This trend has also continued to the type-2 fuzzy systems area, which has been accounted for with the review of papers presented in the previous chapters. In the case of designing type-2 fuzzy systems the problem is more complicated due to the higher number of parameters to consider, making it of upmost importance the use of bio-inspired optimization techniques for achieving the optimal designs of this sort of systems. In this chapter a summary of the total number of papers published in the area of type-2 fuzzy system optimization is presented, so that the increasing trend occurring in this area can be better appreciated. Also, the distribution of papers according to the used optimization technique is presented, so that a general idea of how these different techniques are contributing to the automatic design of optimal type-2 fuzzy systems is obtained.

Figure 10.1 shows the total number of papers published per year describing the application of optimization methods for designing type-2 fuzzy systems in the areas of control, pattern recognition, classification, and time series prediction. From Fig. 10.1 it can be noted that the number of papers published have been increasing each year (in 2011 there appears to be a decline because the information of this year is not complete at the moment of preparing the paper). It is expected that this increasing trend will continue in the future because type-2 fuzzy systems have been recently used more frequently in the applications (and are becoming more popular), and this will require designing more complex type-2 fuzzy systems, which in turn will need even better optimization techniques to achieve solutions more efficiently. It is also worth mentioning that at the moment most of the type-2 fuzzy systems considered in the applications only use interval type-2 fuzzy

O. Castillo and P. Melin, *Recent Advances in Interval Type-2 Fuzzy Systems*, SpringerBriefs in Computational Intelligence, DOI: 10.1007/978-3-642-28956-9_10,

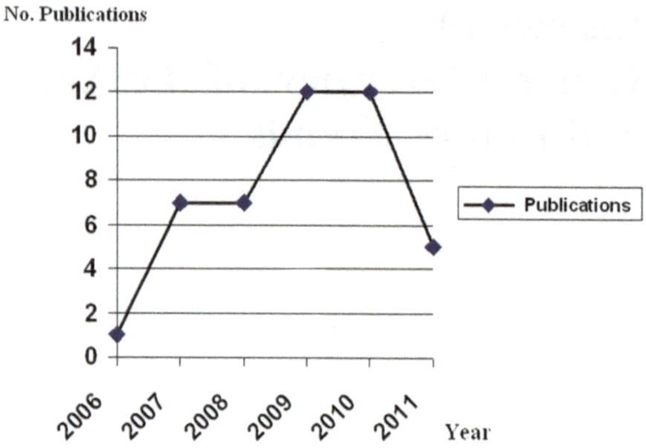

Fig. 10.1 Total publications per year for the 2006–2011 periodof time

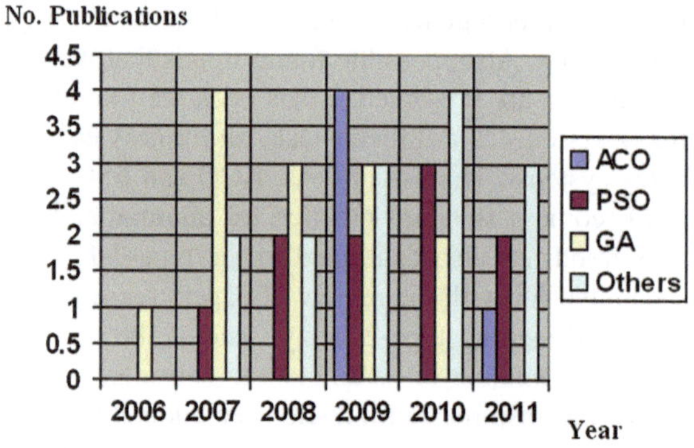

Fig. 10.2 Distribution of publications per area and year

sets due to the higher degree of difficulty in managing and processing generalized type-2 fuzzy sets, but when these generalized type-2 fuzzy sets become more of a standard the design problem would require even more powerful optimization techniques.

Figure 10.2 shows the distribution of the published papers in optimizing type-2 fuzzy systems according to the different bio-inspired optimization techniques previously mentioned. From Fig. 10.2 it can be noted that the use of GAs have been decreasing recently, on the other hand the use of PSO, ACO and other methods have been increasing. The reason for the increase in use of PSO and ACO may be due to recent works in which either PSO or ACO have been able to outperform GAs for different applications. Regarding the question of which method would be the most appropriate for optimizing type-2 fuzzy systems, there is no easy answer. At the moment, what we can be sure of is that the techniques mentioned in this paper and probably newer ones that may appear in the future, would certainly be tested in the optimization of type-2 fuzzy systems because the problem of designing automatically these types of systems is complex enough to require their use.

There are other bio-inspired or nature-inspired techniques that at the moment have not been applied to the optimization of type-2 fuzzy systems that may be worth mentioning. For example, membrane computing, harmony computing, electromagnetism based computing, and other similar approaches have not been applied (to the moment) in the optimization of type-2 fuzzy systems. It is expected that these approaches and similar ones could be applied in the near future in the area of type-2 fuzzy system optimization. Of course, as new bio-inspired and nature-inspired optimization methods are being proposed at any time in this fruitful area of research, it is expected that newer optimization techniques would also be tried in the near future in the automatic design of optimal type-2 fuzzy systems.

Index

O. Castillo and P. Melin, *Recent Advances in Interval Type-2 Fuzzy Systems*,
SpringerBriefs in Computational Intelligence, DOI: 10.1007/978-3-642-28956-9,

immobilized whole cells are compared with those of immobilized purified enzymes.

The possibilities of genetic manipulation are also discussed. Already strain improvement in the Pruteen process has been brought about by the introduction of the glutamate dehydrogenase gene (from *Escherichia coli*) to replace the energy-consuming glutamine synthetase/glutamate synthase that normally operates in *Methylophilus methylotrophus*. Theoretically, by introducing genes for key methylotrophic enzymes into other organisms, it might be possible to replace carbohydrate by methanol as a growth substrate for various industrial processes.

References

This list is much more extensive than those in preceding chapters. This is partly because of the diffuse nature of the literature, but is also an indication of the importance of the topic.

Articles in Periodicals

ANTHONY, C. (1978). The prediction of growth yields in methylotrophs. *Journal of General Microbiology* **104**, 91–104.

DROZD, J.W. and WREN, S.J. (1980). Growth energetics in the production of bacterial single-cell protein from methanol. *Biotechnology and Bioengineering* **22**, 353–362.

GAUTHIER, F. and BONEWALD, R. (1980). The use of plasmid R1162 and derivatives for gene cloning in the methanol-utilizing *Pseudomonas* AM1. *Molecular and General Genetics* **178**, 375–380.

GOLDBERG, I. (1977). Production of SCP from methanol—Yield factors. *Process Biochemistry* **12**, (9), 12–18.

KRUG, E.L., LIM, H.C. and TSAO, G.T. (1979). Single-cell protein from C_1 compounds. *Annual Reports on Fermentation Processes* **3**, 141–195.

LASKIN, A.I. (1977). Single-cell protein. *Annual Reports on Fermentation Processes* **1**, 151–180.

MESSING, R.A. (1980). Immobilized microbes. *Annual Reports on Fermentation Processes* **4**, 106–121.

MEYER, O. (1980). Using carbon monoxide to produce single-cell protein. *BioScience* **30**, 405–407.

TANI, Y. and YAMADA, H. (1980). Microbial utilization of C_1 compounds. *Biotechnology and Bioengineering* **22**, (Suppl. 1) 163–175.

WINDASS, J.D., WORSEY, M.J., PIOLI, E.M., PIOLI, D., BARTH, P.T., ATHERTON, K.T., DART, E.C., BYROM, D., POWELL, K. and SENIOR, P.J. (1980). Improved conversion of methanol to single-cell protein by *Methylophilus methylotrophus*. *Nature* **287**, 396–401

Articles in Books

HARDER, W. and VAN DIJKEN, J.P. (1976). Theoretical considerations on the relation between energy production and growth of methane-utilizing bacteria. In *Microbial Production and Utilization of Gases* (H_2, CH_4, CO). pp. 403–418. Edited by H.G. Schlegel, G. Gottschalk and N. Pfennig. Akademie der Wissenschaften, Göttingen.

NORRIS, J.R. (1981). Single-cell protein production. In *Essays in Applied Microbiology* pp. 6/1–6/31. Edited by J.R. Norris and M.H. Richmond. J. Wiley, Chichester.

STOUTHAMER, A.H. (1979). The search for correlation between theoretical and experimental growth yields. In *Microbial Biochemistry* (International Review of Biochemistry, vol. 21) pp. 1–47. Edited by J.R. Quayle. University Park Press, Baltimore.

The following references are all to chapters in *Hydrocarbons in Biotechnology*. Edited by D.E.F. Harrison, I.J. Higgins and R. Watkinson. Heyden & Son, London.

DALTON, H. (1980). Transformations by methane mono-oxygenase. pp. 85–97.

DROZD, J.W. (1980). Whole cell transformations. pp. 75–83.

HAMER, G. and HARRISON, D.E.F. (1980). Single-cell protein: the technology, economics and future potential. pp. 59–73.

HIGGINS, I.J., HAMMOND, R.C., PLOTKIN, E., HILL, H.A.O., UOSAKI, K., EDDOWES, M.J. and CASS, A.E.G. (1980). Electroenzymology and biofuel cells. pp. 181–193.

The following references are all to chapters in *Anaerobic Digestion*. Edited by D.A. Stafford, B.I. Wheatley and D.E. Hughes. Applied Science Publishers, London.

HAYES, T.D., JEWELL, W.J., DELL'ORTO, S., FANFONI, K.J., LEUSCHNER, A.P. and SHERMAN, D.F. (1980). Anaerobic digestion of cattle manure. pp. 255–288.

HOBSON, P.N., BOUSFIELD, S., SUMMERS, R. and MILLS, P.J. (1980). Anaerobic digestion of piggery and poultry wastes. pp. 237–253.

MOSEY, F.E. (1980). Sewage treatment using anaerobic digestion. pp. 205–235.

PFEFFER, J.T. (1980). Domestic refuse as a feed for digesters. pp. 187–203.

PYLE, D.L. (1980). Anaerobic digester designs in the third world. pp. 345–376.

STEWART, D.J. (1980). Energy crops to methane. pp. 303–319.

WHEATLEY, B.I. (1980). The gaseous products of anaerobic digestion—biogas. pp. 415–428.

The following references are all to chapters in *Microbial Growth on C_1 Compounds*. Proceedings of Third International Symposium, Sheffield, 1980. Edited by H. Dalton. Heyden & Son, London.

HARDER, W., VAN DIJKEN, J.P. and ROELS, J.A. (1981). Utilization of energy in methylotrophs. pp. 258–269.

FAUST, U., PRÄVE, P. and SCHLINGMANN, M. (1981). Single-cell protein from methanol: production for a high quality product. pp. 335–341.

SMITH, S.R.L. (1981). Some aspects of ICI's single-cell protein process. pp. 342–348.

Overall summary

In addition to the traditional circulation of carbon via multicarbon compounds in the biosphere ($CO_2 \rightarrow$ green plants $\rightarrow$ animals $\rightarrow CO_2$; death and decay $\rightarrow$ non-living organic compounds $\rightarrow CO_2$) a large quantity of carbon also circulates through methane and reduced C_1 compounds. This has only been fully appreciated in recent years since accurate data for the methane content of the atmosphere have been available. Cycles of the type $CO_2 \rightarrow$ methanogenic bacteria $\rightarrow CH_4 \rightarrow$ methylotrophic bacteria $\rightarrow CO_2$; death and decay $\rightarrow$ non-living organic matter $\rightarrow CO_2$ and analogous systems involving methanol and carbon monoxide exist as what we call the methane cycle. The abundant methane in this cycle is produced by anaerobic bacteria called *methanogenic bacteria*. Some seven genera of methanogens exist: they are thought to be very ancient forms of life, and the genera are taxonomically greatly different. These bacteria live in three main types of habitat: in anaerobic sediments in lakes, paddy fields, tundra or anaerobic sludge digesters; in the rumen of cattle, sheep and similar mammals; and in geothermal springs. In these ecosystems we have a mixed bacterial population that breaks down organic compounds anaerobically to methane and carbon dioxide. Whether other organic compounds are also formed depends on the ecosystem. Thus, in the rumen volatile fatty acids are formed that are used as an energy source by the ruminant, and the gas produced thus contains less methane and more carbon dioxide. In other ecosystems, the final products are often merely methane and CO_2 in a ratio of 2:1.

The methane is formed by the eight-electron reduction of CO_2 by hydrogen gas, catalysed by the methanogenic bacteria. The standard free energy change for this reaction, $-136\,kJ\,mol^{-1}$, is sufficient for the methanogen to obtain energy for growth. The hydrogen and CO_2 are produced by other bacteria in the ecosystem, and the removal of hydrogen by the methanogens has a marked effect on both the growth and fermentation products of the other bacteria. This is called inter-species hydrogen transfer. Methane is formed by the reduction of bound C_1 units, probably as thiolesters of coenzyme M($HSCH_2CH_2SO_3^-$). The reduction process involves several novel coenzymes and results in ATP formation in intracellular membranes. Many details of the biochemistry of this process are not yet understood. The pathway of carbon assimilation into cell material involves the conversion of two molecules of CO_2 into acetyl-coenzyme A by an unknown route not involving acetate. Acetyl-coenzyme A is then carboxylated successively to C_3 and C_4 derivatives from which the cell constituents are formed. The methanogens are distinguished from the eubacteria by a series of unusual cell components (Table 2.2). Their cell walls are different and their metabolism involves the use of novel coenzymes, which are significantly different from those in eubacteria—they contain deazaflavins, nickel–porphyrin compounds and novel pteridines—as well as the more usual flavins and cytochromes.

The methane formed by methanogenic bacteria can be used as a growth substrate by certain groups of aerobic bacteria. In addition to methane, a wide range of other reduced C_1 compounds such as methanol, carbon monoxide, formate, methylated amines and methylated sulphur compounds can often also support growth of these *methylotrophic bacteria*. Methylotrophs recycle the methane formed by methanogens by converting it to cell material (biomass) and CO_2. (There is also a significant chemical transformation of methane to methanol, formaldehyde and carbon monoxide in the upper atmosphere. Some of this material comes back to earth and is available to methylotrophic bacteria). Methylotrophs can be obligate, in which case they can only grow on reduced C_1 compounds, or facultative. Facultative methylotrophs are either heterotrophic methylotrophs, which as an alternative to C_1 compounds can also use multicarbon compounds, or autotrophic methylotrophs. The latter, as an alternative to methylotrophic growth, can also grow photosynthetically or chemoautotrophically and fix CO_2. Three pathways of carbon assimilation are known in methylotrophic bacteria. The ribulose bisphosphate cycle or Calvin cycle of CO_2 fixation operates in autotrophic methylotrophy. Heterotrophic methylotrophs use either the hexulose phosphate cycle of formaldehyde fixation, or the serine pathway of carbon assimilation. In the latter carbon is assimilated both as formaldehyde and as CO_2.

Energy is obtained in methylotrophic bacteria by the oxidation of the C_1 substrate to CO_2. Generally speaking, the tricarboxylic acid cycle does not fulfil any oxidative role, and indeed probably does not function in obligate methylotrophs. Sometimes the C_1 oxidation pathway is a linear sequence: Methane $\rightarrow$ Methanol $\rightarrow$ Formaldehyde $\rightarrow$ Formate $\rightarrow$ CO_2 or $(CH_3)_nN \rightarrow$ Formaldehyde $\rightarrow$ etc., but with some organisms using the hexulose phosphate cycle there are also present in the cells enzymes that make up a dissimilatory hexulose phosphate cycle that oxidizes formaldehyde to CO_2 via sugar phosphates rather than via formate. How significant this cycle is has not yet been established for certain. Methane is converted to methanol by two kinds of hydroxylation system which are non-specific and can hydroxylate various other compounds. Methanol is oxidized in bacteria by methanol dehydrogenase, a novel enzyme with a quinone prosthetic group called methoxatin or pyrrolo-quinoline quinone. This enzyme transfers electrons from methanol to feed into the electron transport chain at the level of cytochrome *c*. This means that only one molecule of ATP can be formed per molecule of methanol oxidized to formaldehyde. Methanol dehydrogenase lies on the periplasmic surface of the cell membrane. Cytochrome *c* plays an essential role in methylotrophic growth, but it is not yet certain that this is merely because it accepts electrons from the prosthetic group of methanol dehydrogenase. Methylated amines are oxidized via alternative pathways involving either mono-oxygenases or dehydrogenases. Possession of the latter enables these amines to be used as carbon source during denitrifying growth in the total absence of oxygen. At least three mechanisms for the oxidation of methylamine have been described.

Control is important in methylotrophic metabolism, because the largely separate assimilatory and dissimilatory pathways have clear-cut branch points, where the two pathways diverge. Control of enzyme activity is essential to divert metabolites into the correct pathway according to the physiological needs and energy status of the cell. Because many of the enzymes of methylotrophic metabolism play no other role in the cell, control at the level of enzyme formation is also necessary, so that during the heterotrophic or autotrophic growth of facultative methylotrophs, the 'methylotrophic enzymes' can be repressed.

The only eukaryotes studied that are capable of methylotrophic growth are a relatively small number of yeast species, particularly members of the genera *Candida*, *Hansenula*, *Pichia* and *Torulopsis*. Both the assimilatory and dissimilatory pathways show significant differences from bacteria. Dissimilation of methanol involves its oxidation to formaldehyde by an oxidase, which does not generate energy. The formaldehyde is oxidized to CO_2 by extramitochondrial enzymes that generate NADH, which on reoxidation produces less ATP than intramitochondrial NADH. Assimilation proceeds via the dihydroxyacetone pathway, a newly characterized route in which the key enzyme is a special type of transketolase.

The two major applications at the industrial level of micro-organisms described in this book are the use of methanogenic bacteria in the anaerobic digestion of domestic, sanitary, industrial and agricultural waste, and the use of methylotrophic bacteria for single-cell protein production. (Single-cell protein is protein from microbial sources used as food for vertebrates.)

Anaerobic digestion is widely used in the disposal of sewage and of waste from food-processing factories. Applications have also been suggested for the disposal of cattle, poultry or piggery wastes without spreading obnoxious smells. The digestion process turns animal excrement into a relatively odourless sludge (which can be used as a fertilizer) with the generation of biogas (67% methane, 33% CO_2), which can be used for heating purposes or for generating electricity. It is likely that in the future most large farms will become energy-sufficient by the use of biogas.

Methanol has now become an important source of single-cell protein. Yeasts can be used for this, but are not as efficient as bacteria in terms of energy yield. Methanol has several advantages as a carbon source over methane, the major one being its water solubility. ICI Ltd are now producing 40 000 tonnes per year of broken, dried bacterial cells of the hexulose phosphate cycle bacterium *Methylophilus methylotrophus*. The product, called Pruteen, is used mainly as a substitute for soyabean- and fish-meal in poultry feeds.

There are also several possible applications that are not yet beyond the laboratory stage, including biotransformations using methylotrophic bacteria with particular reference to methane mono-oxygenase, the use of immobilized cells and enzymes and the possibilities of improving the industrial efficiency of micro-organisms and their adaptation to new uses, by genetic manipulation.

Further Reading

ANTHONY, C. (1982) *The Biochemistry of Methylotrophs*. Academic Press, London.

DALTON, H. (Editor) (1981). *Microbial Growth on C_1 Compounds*. Proceedings of the third International Symposium, Sheffield, 1980. Heyden & Son, London.

HARRISON, D.E.F., HIGGINS, I.J. and WATKINSON, R. (1980) *Hydrocarbons in Biotechnology*. Published on behalf of the Institute of Petroleum. Heyden & Son, London.

SCIENTIFIC AMERICAN (1981). Issue on *Industrial Microbiology* **245** (3), 42–170.

STAFFORD, D.A., WHEATLEY, B.I. and HUGHES, D.E. (Editors) (1980). *Anaerobic Digestion*. Applied Science Publishers, London.

TANNENBAUM, S.R. and WANG, D.I.C. (Editors) (1975). *Single-cell Protein II*. MIT Press, Cambridge, Massachusetts.

TAYLOR, G.T. (1982) The methanogenic bacteria. *Progress in Industrial Microbiology* **16**, 231–329.

ZEIKUS, J.G. (1980). Chemical and fuel production by anaerobic bacteria. *Annual Review of Microbiology* **34**, 423–464.

Glossary

This list does not include terms that are defined in the text.

Atractyloside. A toxic glycoside isolated from thistles, which inhibits the transfer of adenyl nucleotides through the mitochondrial membrane.

Biosphere. That component of the earth's surface which comprises living organisms and their products.

Chromophore. The chemical grouping in a molecule responsible for its light-absorbing properties and hence its colour.

N,N′-**Dicyclohexylcarbodiimide** $C_{13}H_{22}N_2$. Very toxic reactive chemical inhibiting ATP synthesis in mitochondria.

Ferredoxin. A family of nonhaem iron–sulphur proteins functional as electron carriers found mainly but not exclusively in anaerobic bacteria. E_0' about $-430\,mV$.

Food chain. Relationship between living organisms A, B, C, in which A eats B, which in turn has eaten C etc. Sometimes the components of a food chain may be products of living organisms, such as milk, or dead bodies, etc.

Muramic acid. 2-Amino-3-*O*-(lactyl)-2-deoxy-D-glucose $C_9H_{17}NO_7$, one of the two amino sugars making up the carbohydrate backbone of murein (see below).

Murein. The major component of the eubacterial cell wall consisting of a polysaccharide backbone of alternating *N*-acetylmuramic acid and *N*-acetylglucosamine cross-linked by peptide chains containing D- and L-amino acids attached to the muramic acid. Also called peptidoglycan.

Nigericin. Macrotetralide antibiotic that allows cations to be conducted through membranes.

Pectin. Polymer found in the roots, stems and fruits of plants, molecular weight 20 000 to 400 000, mainly polygalacturonic acid with up to 60% of the carboxyl groups esterified with methanol.

Pentosan. Polymer of pentoses, particularly of xylose. Xylans are the major component of the *hemicellulose* fraction of the plant cell wall. Hemicelluloses are non-cellulose, non-lignin cell wall components.

Peroxisome. Intracellular organelle of eukaryotes with single membrane boundary containing oxidases such as urate, glucose or D-amino acid oxidase, and catalase. Also called microbody.

Silage. Green grass and vegetable crops stored damp and anaerobically in a pit. Primary fermentation (see Chapter 2) of starch and cellulose occurs, giving a product rich in fatty acids.

Squalene. Isoprenoid hydrocarbon $C_{30}H_{50}$ found in many natural oils. Biosynthetic precursor of sterols.

Index